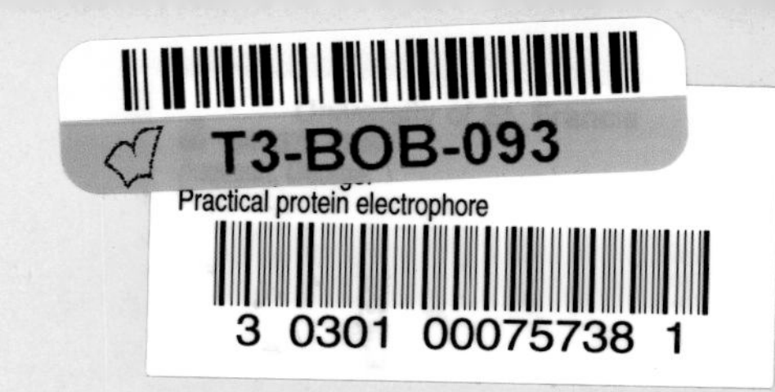

Practical Protein Electrophoresis for Genetic Research

Practical Protein Electrophoresis for Genetic Research

George Acquaah, Ph.D.
Langston University
Langston, Oklahoma

DIOSCORIDES PRESS
Theodore R. Dudley, Ph.D., General Editor
Portland, Oregon

Disclaimer

Mention of a trademark, proprietary product, or vendor does not constitute a guarantee or warranty of the product by the publisher or author and does not imply its approval to the exclusion of other products or vendors. Furthermore, the author does not guarantee that any of the protocols described will work in all situations even though they are accurate to the best of his knowledge.

ISBN 0-931146-22-4
Printed in Hong Kong

DIOSCORIDES PRESS
9999 S.W. Wilshire, Suite 124
Portland, Oregon 97225

Library of Congress Cataloging-in-Publication Data

Acquaah, George.
Practical protein electrophoresis for genetic research / George Acquaah.
p. cm.
Includes bibliographical references and index.
ISBN 0-931146-22-4
1. Proteins--Separation. 2. Isoenzymes--Separation. 3. Enzymes--Separation. 4. Gel electrophoresis--Technique. 5. Molecular genetics--Technique. 6. Isoenzymes--genetics--laboratory manuals. I. Title.
[DNLM: 1. Electrophoresis, Starch Gel--methods--laboratory manuals. QU 25 A186p]
QP551.A27 1992
574.19'245--dc20
DNLM/DLC
for Library of Congress 92-4755
CIP

CONTENTS

To the "home team": Theresa (quarterback), Parry (running back), Paa Kwasi (defensive tackle), and Lady Bozuma (homecoming queen and cheerleader). In my book you will always be winners!

FOREWORD

Gel electrophoresis of proteins has become a standard and extremely powerful tool in many areas of plant and animal biology. Isozyme electrophoresis in particular has positively influenced a large number of biological disciplines. Isozymes, originally defined as enzyme variants present in the same individual having identical or similar functions, have become routine sources of data in population biology, evolutionary biology, systematics, agronomy, and forestry.

Despite the great importance of gel electrophoresis of proteins in plant biology, few attempts have been made to integrate into one volume discussion of the principles of protein electrophoresis, methodological nuances, gel interpretation, troubleshooting, isozyme nomenclature, data collection and analysis, and genetic and statistical procedures, as well as provide a summary of examples in genetic research. Dr. Acquaah does an excellent job of accomplishing this integration. His book spans a range of topics not covered previously in a single volume. It is an excellent primer for the novice initiating protein electrophoretic investigations and is also useful for the experienced investigator.

The volume focuses on isozyme electrophoresis. The author discusses the principles of isozyme electrophoresis and histochemical staining and the many reasons for the popularity of electrophoretic data, the foremost of which is that isozymes provide a series of easily scored, single-gene markers. Importantly, the disadvantages of this approach, such as the limited number of markers generated and the representativeness of isozymes in terms of the entire nuclear genome, are also noted.

Perhaps the greatest strength of this volume is the attention given to methodological detail. Dr. Acquaah provides invaluable accounts of both starch gel electrophoresis (SGE) and polyacrylamide gel electrophoresis (PAGE). The discussions of both approaches are extremely detailed, yet easily read and understood. Not only are basic principles provided, but excellent descriptions of methods, the construction of apparati, and other critical data are given. The discussions of methodology eliminate much of the mystery behind protein electrophoresis and are a must for all beginners. For example, the detailed discussion of SGE includes actual plans for the construction of wick vs. wickless electrophoretic apparati; a list of commonly employed gel and electrode buffer combinations and extraction buffers; important advice on sample collection and storage; details of sample preparation, gel pouring, and slicing; actual plans for gel slicers; methods of gel fixation; and design strategies for loading samples onto starch gels for easy comparison among new samples and controls. All of the methods and equipment are excellently illustrated.

Similarly, the account of PAGE is also thorough. Using a series of photographs and illustrations, the author discusses the different types of vertical PAGE apparati, provides plans for construction, and outlines important methods, including gel casting, sample loading, gel staining, and destaining. He provides selected protocols for gel preparation and staining and presents specialized applications of PAGE, including isolectric focusing and two-dimensional electrophoresis.

Other chapters of the volume concentrate on troubleshooting for both SGE and PAGE. Dr. Acquaah notes the commonly encountered methodological problems and outlines the best approaches for discovering and overcoming them.

Another extremely important contribution of this volume is the chapter on setting up an isozyme laboratory. The necessary materials and chemicals are listed, as are potential suppliers. Assembling this information has always been difficult for the novice.

Authors of many other methodologically oriented reviews have considered their jobs complete after instructing the novice in the construction of gel rigs, gel pouring, and choice of appropriate electrophoretic buffers and staining protocols. Another real asset of this volume, however, is that it provides the beginner with helpful information in data collection and analysis. Other features that facilitate the development of the beginner's skills are the examples of data sheets and the discussions and illustration of recording of electrophoretic data. Discussions of the principles of isozyme data interpretation, the possibility of detecting ghost banding and other artifactual bands, and the basics of isozyme nomenclature give the electrophoretic novice critical information that previously had to be obtained through word of mouth or personal experience.

In the past, after isozyme data were obtained, the electrophoretic novice was left to seek basic population genetic texts (or experienced veterans of isozyme electrophoresis) to apply appropriate genetic calculation and statistical tests to the data set. Dr. Acquaah provides extremely helpful "worked examples" based on actual isozyme data sets. These data sets demonstrate the calculation of allele frequencies, genetic polymorphism, and heterozygosity (both observed and expected). Also explained are tests of an independent assortment of alleles and linkage between an isozyme locus and a locus controlling a morphological trait.

Dr. Acquaah does not pretend to provide a source of detailed information regarding the application of isozyme electrophoresis to fields such as population biology, systematics, and crop breeding. Nonetheless, general references in these areas are provided to facilitate the beginner's successful entry into the appropriate literature.

The author is to be congratulated for combining in one volume information regarding isozyme methodology and troubleshooting, gel scoring and interpretation, and basic genetic computations. This book will greatly expedite the learning process for any individual initiating an isozyme electrophoretic investigation.

Dr. D. E. Soltis
Department of Botany
Washington State University
Pullman, Washington 99164-4230

PREFACE

This book was written to enable the researcher who is not familiar with electrophoresis to use the technique in research with little or no additional help. The style is a practical cookbook format, describing all steps in detail, supported with photographs and diagrams. More importantly, substantial theoretical information explains some of the basic principles of electrophoresis and the concept of isozymes.

The occasional and especially the uninitiated user of the technique of electrophoresis, who may not have had time to survey the literature extensively, will benefit immensely from the general information in this book. It should be helpful to both students and experienced users of the technique. An extensive list of pertinent literature has been provided for further detailed consultation. A broad range of buffer types and staining protocols and a detailed list of equipment and chemicals needed for electrophoresis have been provided. The protocols are applicable to both plant and animal studies, with or without modification.

By using this book, the researcher will be able to assemble the basic apparatus for electrophoresis and to conduct, analyze, and interpret electrophoretic data. Most of the examples given have been drawn from the experiences of the author working with soybean, dry beans, and sugar beet. Other resource materials have been duely acknowledged, where possible.

Starch gel electrophoresis (SGE) is a very simple yet versatile technique for biological research. It is not very demanding in terms of accuracy and precision. The quantities of many reagents need only be estimated, and protocols need not be strictly followed in some cases. Many of the modifications to original protocols were developed by trial and error. The technique allows room for flexibility, creativity, and experimentation. This is not to say that standards do not exist in the application of the technique, but simply that it is more important to understand the "whys" of procedures so that the "hows" can be intelligently modified to achieve the desired results.

The uninitiated should not be surprised to find variation in procedures from one laboratory to another. Some of this variation stems from the diversity in species and tissues used as samples for electrophoretic analysis. The same enzyme may behave differently under different situations, and therefore generalizing protocols is not always safe. Although picking up ideas from other laboratories can be helpful, one should not abandon a tried and proven technique for the sake of novelty. "If it ain't broke, don't fix it!"

I hope that this work helps users in their efforts to contribute knowledge to their field of science.

ACKNOWLEDGMENTS

I am greatly indebted to the following persons associated with the Michigan State University for their very useful comments, suggestions, encouragement, and time at various stages of the preparation of this book: Clay Sneller, Dr. J. Hancock, Jr., Dr. G. Hosfield, Dr. S. Krebs, and Dr. E. Watts. Their input is deeply appreciated. Clay Sneller deserves very special mention for his critical role in my laboratory work and experience. The financial support and employment opportunities offered by Dr. J. W. Saunders and Dr. T. G. Isleib made this book possible. The experience acquired in their laboratories provided many of the examples cited in this book. For this and their general kindness, cooperation, and intellectual stimulation, I express my heartfelt appreciation and thanks.

I also deeply appreciate the technical help and guidance I received from Dr. L. Copeland and Dr. S. O. Acquaah. Special thanks go to Mrs. L. Ngige and especially Ms. S. Copenhagen for editorial assistance.

Without the support of my family, who endured my long and frequent absence from home, what is now presented in black and white would have remained mere "head knowledge." The least I can do is to dedicate this work to the "home team."

Chapter 1

GENETIC PRINCIPLES ASSOCIATED WITH ISOZYMES

This chapter introduces the fundamental concepts of isozyme genetics. The mechanisms by which isozymes arise and the genetics underlying such mechanisms are briefly explained. The controversial issue of the neutrality of isozyme variation in natural populations also is discussed.

1.1. Polymorphism

Polymorphism may be defined as the simultaneous occurrence within or between populations of multiple phenotypic forms of a trait attributable to the alleles of a single gene or the homologs of a single chromosome (Suzuki et al. 1981). A classic example is the ABO blood group polymorphism in humans. Recurrent mutations of genes produce variability in a natural population. Although some loci are variable in the sense described above (polymorphic), others are nonvariant (monomorphic) (Figure 1–1). The question of how much allelic variation is required at a locus to qualify it as polymorphic is addressed below.

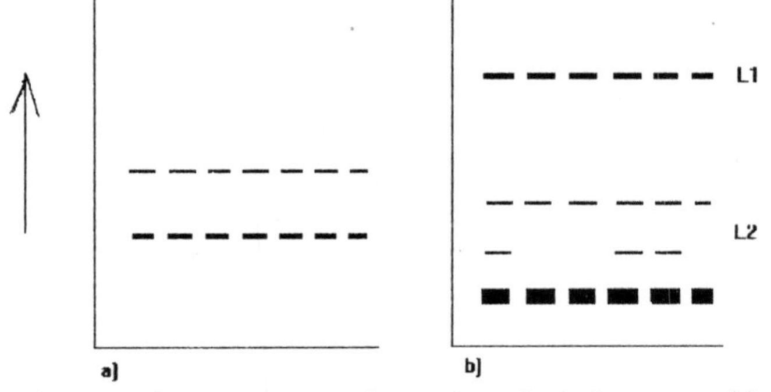

Figure 1-1. Isozyme patterns for two hypothetical enzymes: (*a*) a monomorphic enzyme and (*b*) an enzyme that is monomorphic at one locus (L1) but polymorphic at another (L2).

1.2. The Isozyme Concept

Markert and Moller (1959) first coined the term *isozymes* (from *isoenzymes*) to describe the different molecular forms of an enzyme in a species that share a common catalytic activity. Since then, other, more restrictive, definitions have been proposed. Brewer and Sing (1970) defined isozymes as the multiple molecular forms of enzymes derived from the same organism (or tissue culture) that share a common catalytic activity. The Markert and Moller (1959) definition is accepted by the International Union of Biochemistry (IUB) (1979) Commission on Biological Nomenclature. The salient points of the concept of isozymes as perceived by researchers and reviewed by Markert (1977) are summarized below:

1. Multiple molecular forms of enzymes (isozymes) are common in organisms (in fact, they are ubiquitous).
2. Isozymes share a common catalytic activity. Each isozyme has a specific role in the metabolic pathway and functions in harmony with other enzymes within the organizational framework of cells.
3. Isozymes often exhibit tissue or cell specificity.
4. Molecular heterogeneity of enzymes confers flexibility, versatility, and precision upon an organism in terms of metabolic functions.
5. Molecular multiplicity is desirable for biological efficiency.

Since every gene can mutate, it is reasonable to assume that every enzyme can exist as an allelic isozyme, at least occasionally (Markert 1977).

1.3. Origin of Isozymes and Isozyme Systems

Isozymes arise in nature as a result of two general mechanisms, namely, genetic and epigenetic. The origin of gene multiplicity is often attributable to the duplication of the gene coding for an enzyme with subsequent divergence through mutation to produce different enzymes; polyploidization (a source of gene duplication) of the entire genome or part of it also has been implicated (MacDonald and Brewbaker 1974; Markert 1977; Baum and Scandalios 1982; Gottlieb 1982). These chromosomal and genic aberrations constitute the genetic origin of isozymes in an organism. In the process of translating the genetic message encoded in the DNA, the resulting polypeptide is subject to physical and chemical alterations. The above events constitute a second source (epigenetic) of isozymes (Munkres and Richards 1965; Markert 1977). Epigenetically derived enzyme forms are not, however, considered isozymes by some researchers (Brewer and Sing 1970; Moss 1982). Isozymes derived from the two mechanisms may further be classified under systems for generating isozymes, as was done by Markert and Whitt (1968) and Markert (1977), whose work I have synthesized and reorganized as follows:

1.3.1. Genetic Mechanisms

1.3.1.1. Multilocus System I The multilocus system I is the simplest mechanism by which isozymes arise. The distinguishing characteristic of the multilocus system I is that different genes code for independent proteins with the same enzymatic activity. The various genes are nuclear in origin but their protein products are located in different parts of the cell, one in the cytosol and the other in an organelle (chloroplast, mitochondria). Enzymes in this category are functionally similar (Moss 1982). An example of the system I is the malate dehydrogenase enzyme.

1.3.1.2. Multilocus System II The multilocus system II is genetically similar to system I except that the enzymes involved are polymeric (multiple polypeptide chains) and the subunits are encoded by more than one locus. The capacity for the formation of homomultimers (identical multiple peptide chains) and heteromultimers (nonidentical multiple peptide chains or hybrid isozymes) exists in system II. It must be borne in mind that heterozygous organisms can produce isozymes that simulate the multilocus systems. An example of system II is the lactate dehydrogenase enzyme. The subject of multimerism is discussed further below under Quaternary Structure of Isozymes (1.5).

1.3.1.3 Unilocus-polymeric System Enzymes in the unilocus-polymeric system display a series of polymers that consist of identical subunits. If an enzyme has allelic genes at its locus (or undergoes mutation), it would exhibit the combined characteristics of the multilocus system II and the allozyme system (described below). An example of an enzyme in this category is glutamate dehydrogenase.

1.3.1.4. Allozyme System The term *allozyme* was coined by Prakash et al. (1969) to describe isozymes encoded by allelic genes (just as alternative alleles at a color locus may code for either white or purple flowers). Alleles at various loci may be modified to produce isozymes that are distributed in a population according to Mendelian laws of inheritance (Moss 1982). If an allozyme is multimeric, both homomers and heteromers are produced in heterozygous individuals.

1.3.2. Epigenetic Mechanisms

Post-translational modifications may be covalent or noncovalent and may involve changes due to processes such as phosphorylation, aggregation, deamination, acylation, partial chain cleavage, and association with other proteins (Moss 1982). Some examples are as follows:

1.3.2.1. Post-translational Addition Post-translational processing of polypeptide chains may result in molecules being attached to or combined with a chain by the process of phosphorylation. Phosphatase isozymes are commonly derived from post-translational sources.

1.3.2.2. Post-translational Deletion During the processing of translation products, terminal fragments may be removed from a polypeptide chain, as in partial proteolysis, which produces trypsin from trypsinogen (Markert 1977). Peptidase isozymes are frequently formed via post-translational deletion processes.

1.3.2.3. Post-translational Conformation The structural configuration of proteins may be differentially modified (e.g. through the folding of polypeptide chains) to produce enzyme heterogeneity. The isozymes resulting from structural modification are called *conformational isozymes* (Munkres and Richards 1965; Motojima and Sakaguchi 1982). Differences in folding of polypeptide chains cause different proportions of charged carboxyl and amino groups to be exposed, resulting in differential net charge and hence differential mobility of proteins in an electric field (Brewer and Sing 1970). If allosteric modifications such as described above are to generate isozymes, they must be stable (Markert 1977).

1.4. Topographical Location of Isozymes in the Cell

Most isozyme systems are located in the cytosol, either in solution or bound to cell membranes (Markert 1977). Weeden (1983a) listed the following locations in an organism where isozyme variation may be found: (*a*) within the same subcellular compartment; (*b*) in different compartments of a cell; (*c*) in different cells of a tissue; or (*d*) at different developmental stages. The minimum number and subcellular location of isozymes of each enzyme in higher plants appears to be highly conserved in evolution (Gottlieb 1982). This conservation reflects the nature of metabolic activities found in different subcellular compartments (Mitton et al. 1979). Not only are isozymes conserved, but the extent of variation in the gene coding for an enzyme is affected by compartmentalization. Isozymes of cytosolic origin are often more variable than those of organellar sources (Gottlieb and Weeden 1981; Ward and Skibinski 1988). Fractionation of crude protein extracts is the customary procedure for ellucidating the number and subcellular location of isozymes in plants (Nishimura and Beevers 1979).

1.5. Quaternary Structure of Isozymes

Proteins can exist in one of four levels of structural complexity, of which the primary structure (amino acid sequence in a polypeptide chain) is the simplest. The most complex form, the quaternary structure, is attained through the folding and aggregation of polypeptide units. When an enzyme comprises one polypeptide chain, it is called a *monomer.* An enzyme comprising aggregates of polypeptide chains is described as a *multimer* (*polymer* or *oligomer*). More specific classification can be assigned based on the number and kinds of polypeptide chains in an aggregate. If two subunits aggregate, the product is called a *dimer.* An aggregation of four subunits constitutes a *tetramer.* An aggregate is also described as either *homo-* or *hetero-* (as in *homodimer* or *heterodimer*), depending on whether the subunits of the aggregate are identical or not. Enzymes with larger monomers are often more polymorphic than those with smaller ones, and monomeric enzymes are often more polymorphic than multimeric enzymes (Kimura 1979). The quaternary form appears highly conserved. When an enzyme occurs as a dimer in one species, it will very likely occur in the same form in others.

1.6. The Isozyme Neutrality Debate

Whether or not allozyme variation in natural populations is adaptively neutral is still under debate (Lewontin 1974). This debate stems from the neutral theory or, more precisely, the "neutral-mutation/random-drift" theory of molecular evolution proposed by Kimura (1968). The theory makes the following claims:

1. Most evolutionary changes at the molecular level are caused by random fixation of selectively neutral mutants (not by natural selection acting on advantageous mutants, as Darwin's theory claims).
2. Most intraspecific variability at the molecular level (protein and DNA inclusive) is neutral (i.e. most polymorphisms are not adaptive).

This theory is not in favor with some selectionists. Perhaps the most concrete evidence for the neutral theory stems from results from DNA/RNA sequencing. For example, nucleotide substitutions within codons that do not result in amino acid changes take place at a much higher rate than substitutions that do alter amino acids (Kimura 1980). Dykhuizen and Hartl (1980) showed that neutral alleles may have some potential for selection under different environments. In other words, neutral mutations could be the raw material for adaptive evolution (Kimura 1980).

Electrophoretic variants may alter the ultimate fitness of individuals (Lewontin 1974; Khaler et al. 1980; Cavener and Clegg 1981). Lewontin (1974) and Nei (1975) demonstrated that isozyme variation may be adaptive and indirect. The significance of the adaptation of enzyme polymorphism in physiological (Marshal et al. 1973) and geographic (Brown et al. 1976) studies has been documented. The outcome of studies such as those cited above prompted the recommendation that different distinct environments be considered when sampling genotypes for isozyme research (Brown 1978). Allard et al. (1972) reported that over a period of 25 generations, the allelic frequency of esterases in barley changed significantly. Such changes are not likely to occur in most isozymes over short periods, however, even when selection pressure is directed at the isozyme genes themselves (Nielsen 1985). On the other hand, the isozyme genes may be linked to other genes in the genome that may have some selective advantage, so that while the other genes are being directly selected for, the isozyme genes may be indirectly selected. Evidence has shown that it is not possible to study single loci as though they were completely independent of their immediate neighbors (Marshall and Allard 1970; Clegg et al. 1972).

1.7. Choosing Enzymes for Genetic Research

Enzymes have various physiological functions. Some are substrate-specific whereas others are not; certain enzymes have key roles in central metabolism whereas others are involved in secondary metabolism (Johnson 1974). Different classes of enzymes may have inherently different levels of genetic variation, so that nuclear loci whose protein products function in the cytoplasm will probably have a level of genetic variation different from those whose products function in organelles (Gottlieb and Weeden 1981). Therefore, to sample the variation in a population effectively, enzymes from the different physiological classes should be assayed. As Lewontin (1974) pointed out, for a truly random sample of a larger population, enzymes must be chosen without any bias (no a priori knowledge of their roles).

Very often, however, the researcher has no real choice but is limited by the number of enzymes for which staining protocols have been developed. Furthermore, one researcher may not succeed in using a protocol in a situation even though it worked for others or may not be able to interpret electrophoretic results for an enzyme. For certain studies (e.g. on the diversity of a population), bias due to a limited set of enzymes is unavoidable. The enzyme-staining protocols presented in this book include a fairly wide range of classes of enzymes.

Apart from the constraint of limited staining schedules, other factors also influence the choice of enzymes for research: for example, ease, cost, safety, and reliability of protocols. Some enzymes are easier to assay than others, and staining protocols differ in cost. The researcher may find that certain enzyme assays produce consistent results whereas others are variable and more difficult to understand. Some stains use components that are highly toxic and must be handled with caution. Special equipment may be required for certain assays (e.g. source of ultraviolet light for fluorescent staining schedules).

The question of how representative electrophoretically determined variation is of the variation within a genome, a population, or a species is frequently posed. Crawford (1983) remarked that the question is moot because no taxonomic character can be said to be representative. The enzyme loci that can be routinely investigated for polymorphism are between 10 and 40 for most species. Therefore, a representative, unbiased estimate of the total diversity of a genome of, for example, 20,000 genes is not possible using a handful of markers. To make the best of the situation, a researcher should assay as many enzymes and classes of enzymes as possible.

Chapter 2

PRINCIPLES OF ELECTROPHORESIS

This chapter introduces the technique of electrophoresis and its underlying physical and chemical bases. The outcome of electrophoresis depends on several factors, most of which originate from the way and the conditions under which it is conducted; hence the investigator has considerable control over the outcome of electrophoresis. These factors are discussed so that the investigator may effectively modify various aspects of the technique to achieve specific purposes.

2.1. Types of Electrophoretic Techniques

Electrophoresis is a versatile biochemical technique to detect genetic variation. The technique was introduced by Tiselius in 1937 and has two variations (Mahler and Cordes 1968). The first, called *moving boundary electrophoresis,* has no support medium. Instead, the molecules to be separated are all placed together in one solution in which they move freely. When an electrical gradient is applied, the molecules migrate toward the electrode with a charge opposite to theirs, with the result that the initial single boundary formed by the mixture of molecules is broken into several boundaries according to the relative mobilities of the various components of the mixture. This technique is useful for separating and analyzing complex protein mixtures. In the second technique, called *zone electrophoresis,* variability is determined by separating multiple forms of enzymes in a solid matrix.

Protein molecules migrate in an electric field because they are charged (discussed below in this chapter). The starch gel electrophoresis (SGE) technique was developed by Smithies (1955) and coupled by Hunter and Markert (1957) with histochemical staining (enzyme activity staining) methods to develop the zymogram technique of locating zones of enzyme activity directly in the solid matrix in which separation was effected. Upon staining, a zymogram reveals chromatic (or achromatic) spots (bands) that correspond to locations in the separation medium to which various isozymes migrated based upon their properties and on the experimental conditions. The zymogram patterns are the "fingerprints" of specific enzymes.

2.2. Types of Support Media Used in Electrophoresis

Separation of protein molecules may be accomplished in a variety of solid media, which may be classified into the following two broad groups on the basis of how they affect migrating protein molecules.

1. *Passive media* do not affect the migrating molecules beyond the restrictions imposed via the pH of buffers employed in electrophoresis. They include paper, cellulose acetate, alumina, and silica.
2. *Active media* impose additional restrictions on migration due to their molecular sieving properties. The investigator can alter the porosity of a medium for more effective separation of molecules that may have identical charge densities (charge per unit length of the polypeptide chain) but that differ in size. Examples of media in this category are agarose, starch, and polyacrylamide gels.

The support media mentioned above have advantages and drawbacks. Generally, electrophoretic support media are nonionic or electrically neutral (Brewer and Sing 1970). Agarose produces fragile gels with large pores. Agarose gels are usually run on a horizontal gel apparatus and for short periods. Polyacrylamide gels have high chemical and mechanical stability, and are able to function over a wide pH range (Pierce and Brewbaker 1973). Furthermore, polyacrylamide gels have a high degree of transparency, have no adsorption or electro-osmotic properties, and are insoluble in most solvents (Takacs and Kerese 1984). Gels prepared from polyacrylamide can be run in the vertical position as opposed to the horizontal position characteristically used with starch gels. Acrylamide is a neurotoxin, however, and must be handled with care. Cellulose acetate sheets offer an excellent separation medium for electrophoresis but at a relatively higher cost. The acetate sheets require short periods (less than one hour) of electrophoresis. Starch gels are less expensive and can be sliced more readily than a polyacrylamide gel. A thick starch gel can be prepared and sliced for staining several enzymes from one electrophoretic run. The quality of resolution, though adequate for many purposes, may be lower in a starch gel than in a polyacrylamide gel, for example.

Takacs and Kerese (1984), demonstrating the wide range in differences of resolving power of different electrophoretic support media, reported the following order of separation power of different media (by different methods) of serum proteins: paper < agar/cellulose < starch < polyacrylamide. In choosing a support medium, the researcher should consider the cost, safety, duration of electrophoresis, ease of handling, and special advantages such as the ability to slice the medium for multiple enzyme staining.

2.3. Gel Visualization

The desired outcome of electrophoresis is a successful separation of molecules such that differences among genotypes can be unambiguously detected. The following factors govern this outcome:

1. *Separation* refers to the distance between bands (stained spots on the gel). To minimize errors in recording electrophoretic data, proper band separation is essential. In turn, separation is dependent upon other factors, including pH of buffers, porosity of support media, and duration of electrophoresis.

2. *Resolution* is concerned with the band definition (width and sharpness). Bands may be adequately separated but may have streaks in between them or may occur in large blobs instead of narrow, well-defined zones. The quality of samples and the type of buffers used, among other factors, will determine the resolution of bands upon staining.

2.4. Electrophoresis Buffer Systems

An electrophoresis buffer system is a combination of gel and electrode buffers of specific ionic species composition, ionic strength, and pH buffering capacity. The pH buffering capacity of a buffer is more critical than its composition (O'Malley et al. 1980). In the sugar beet laboratory at the Michigan State University, tris buffers of different molarities and pHs were found to be adequate for use with all the enzymes investigated (W. Doley, personal communication), although I found that alternative buffers were better for certain enzymes.

Buffers are needed in electrophoresis to impart electrical conductivity to the otherwise electrically neutral support media and to keep protein molecules in an active physiological state. For these purposes, buffers are employed as electrolytes and in the extraction of proteins and preparation and staining of gels. The outcome of electrophoresis depends on the appropriate choice of buffers for all four of the steps listed above. Electrophoretic buffers are classified below on how they affect the molecular integrity of proteins and on their ionic properties.

2.4.1. The Effect of Buffers on the Molecular Integrity of Proteins

Buffers may be placed in the following two classes according to how they affect the molecular intergrity of proteins:

1. *Dissociating buffers* have the capacity of denaturing protein molecules to render them physiologically inactive. Denaturation is brought about by chemicals and heat (discussed below under SDS-PAGE gels). The effect of this treatment is to dissociate protein molecules into their component subunits (polypeptides). In this kind of buffer system, the separation of molecules is mainly on the basis of molecular size.

2. *Nondissociating buffers* preserve the molecular integrity of proteins so that they retain their physiological activity. Separation in this system is on the basis of both charge and molecular size.

2.4.2. The Effect of Ionic Properties of Buffers

A gel buffer of low ionic strength promotes faster migration of charged molecules (less time of electrophoresis) and little heating of the gel. On the other hand, a high ionic strength causes slow migration of molecules and more heating of the support medium. High ionic strength tends to produce sharper band resolution, however (Brewer and Sing 1970). Buffer systems are designed such that the ionic strength of the gel buffer is lower than that of the electrolyte in the buffer tank. The researcher may manipulate the ionic strength of buffers until satisfactory results are obtained. The ionic properties and pH of buffers affect the resolution of isozyme patterns. Some enzymes resolve better when certain combinations of buffers (gel and electrode) are used. With practice, a researcher can match enzymes with specific buffers for best results.

Buffers may also be categorized into the following two classes based on ionic properties:

1. *Continuous buffers* have identical ionic species in both gel and electrolyte. The ionic concentration may be different in the gel and the electrolyte (usually less concentrated in the gel), but the pH is usually identical. Continuous buffers in SGE have a tendency to cause streaking in the sample lanes upon staining, a problem that can be reduced by loading smaller amounts of samples for electrophoresis and by dewicking (discussed below) the gels. An example of a continuous buffer is
   Electrode buffer: 0.15 M tris-citrate pH = 6.5
   Gel buffer: 1:3 dilution of electrode buffer pH = 6.5.
2. *Discontinuous buffers* or multiphasic buffers differ in the gel and the electrolyte. The pH also may be different. An example of a discontinuous buffer is
   Electrode buffer: 0.2 M tris-citrate pH = 7.1
   Gel buffer: 0.05 M histidine-HCl pH = 7.5.

In polyacrylamide gel electrophoresis (PAGE), multiphasic buffer systems employ two kinds of gels in one run: a longer, lower gel (*resolving gel*) and an upper, shorter gel (*stacking gel*), which is cast on top of the resolving gel. The stacking gel has larger pores and has the effect of concentrating the samples into very narrow zones (stacks) before they enter the resolving or separating gel (Ornstein 1964). Stacking also enables large quantities of dilute samples to be loaded into the gel; the samples become concentrated in stacks, thus permitting successful electrophoresis of the dilute samples (Hames 1981).

2.4.3. Choosing an Electrophoresis Buffer System

The capacity unambiguously to separate proteins or enzymes is of paramount consideration in choosing a buffer system, but other factors, such as cost, ease of preparation, and safety are important, too. Tris buffer systems, for example, are less expensive than histidine systems. Furthermore, tris buffers give fewer problems associated with microbial contamination than phosphate buffers. Some buffer systems require extra care in their preparation. Morpholine (*N*-(3-aminopropyl)-morpholine) is an irritant and has low toxicity. Some gel and electrode buffers include certain "additives" such as EDTA and sucrose for enhancing resolution of isozymes on a zymogram. Some of these reagents are inhibitors that suppress or eliminate the expression of volunteer bands that tend to complicate zymograms.

2.5. Protein Biochemistry

Proteins are polymers of 20 amino acids (Zubay 1983). Figure 2-1 shows a typical amino acid structure. A polypeptide has a spiral back bone (α-helix), which is the basis of the secondary structure of proteins, and side chains (designated as R), which are made up of amino acid radicals. In the secondary structure, the side chains are exposed on the outside of the spiral and can be ionized, depending upon the ionizing properties of the solution in which the protein is dissolved. The 20 commonly occurring amino acids may be classified into three groups according to their R characteristics (Table 2-1). The uniqueness of amino acids is determined by their R characteristics.

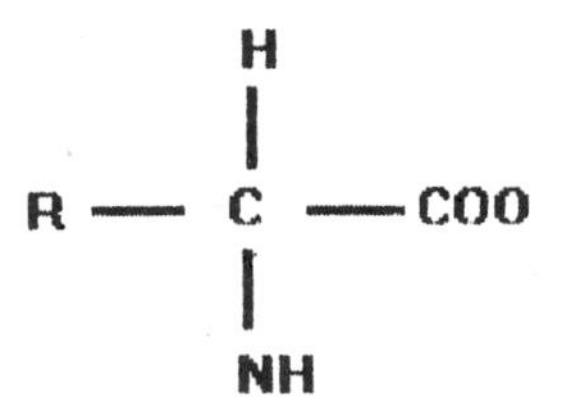

Figure 2-1. A typical structure of an amino acid.

Table 2-1. Classification of the 20 commonly occurring amino acids according to their side-chain (R) characteristics.

Uncharged apolar	Uncharged polar	Charged polar
Alanine	Asparagine	Arginine
Isoleucine	Cysteine	Aspartic acid
Leucine	Glycine	Glutamic acid
Methionine	Glutamine	Histidine
Phenylalanine	Serine	Lysine
Proline	Threonine	
Tryptophan	Tyrosine	
Valine		

Amino acids differ not only in R-group characteristics but also in molecular weight. Different amino acids are linked together in a linear chain by peptide bonds in various combinations and sequences to form specific proteins. A protein may be comprised of amino acids from the three categories in Table 2-1. The net charge of a protein will depend in part on its amino acid composition: If it has more positively charged amino acids such that the sum of the positive charges exceeds the sum of the negative charges, the protein will have an overall positive charge and migrate to the cathode in an electric field. Substitutions in the triplet code can lead to the formation of different amino acids and consequently various proteins of different overall charge. The new protein may be electrophoretically distinguishable from the original one provided that the new one is physiologically active and has an overall charge significantly different from that of the original protein.

As is discussed above in Chapter 1, polypeptide chains are capable of folding and aggregating to produce more complex protein structures, which affect their electrophoretic banding patterns.

2.6. The Role of pH of Solutions

Many amino acids are neutral (pH = 7) with respect to charge, but in that state they are amphoteric (carry both negative and positive charges) and technically are called zwitterions (Zubay 1983). They are comprised of both carboxyl ($-COOH$) and amino ($-NH_2$) subunits, which are acidic and basic, respectively. The pH of the protein environment regulates its net charge. The carboxyl group develops a negative charge through pH-dependent ionization ($-COOH \rightleftharpoons COO^- + H^+$), while the amino group develops a positive charge by the same process ($-NH_2 + H^+ \rightleftharpoons NH_3^+$). These charges even out (net charge equals zero) when a protein is at its isoelectric point (Zubay 1983). Hence, by changing the pH in either direction, the net charge of proteins can be influenced accordingly. The net charge of a protein depends on the number of exposed carboxyl and amino groups that are ionized (Brewer and Sing 1970). The rate of migration in an electric field depends on the magnitude of the net charge of a protein. For unidirectional migration, the pH of an electrophoresis buffer system must be maintained at a constant value to ensure that the original charge is not changed. To achieve this state of stable pH, the solutions used in the gel preparation and as the electrolyte must be properly buffered. Enzymes function best at their specific pH optima, hence staining and other electrophoretic protocols have specified operating pH for different enzymes.

The pH of electrophoretic buffers may be manipulated within a range (depending on the type of buffer) to optimize the resolution of bands of proteins being electrophoresed. The user must manipulate the pH cautiously, however, to maintain the ionic strength of the medium at the lowest possible level to prevent overheating of the gel (Clayton and Tretiak 1972). According to Hames (1981), the further the pH of an electrophoresis buffer is from the isoelectric point of the proteins being separated, the higher the net charge and hence the shorter the duration of electrophoresis. This reduces the diffusion of bands in the gel and enhances the resolution. On the other hand, the closer the pH of the buffer is to the isoelectric point of the proteins, the greater the charge differences between molecules, thus leading to a better separation of molecules (Hames 1981). When a gel is run so that multiple slices can be obtained for staining for different enzymes, there has to be a compromise in the pH to protect the enzymes from physiological damage. It is important to know the pH range over which a protein remains physiologically stable. The isoelectric points of many proteins lie between pH 4 and 7, thus buffers between pH 8 and 9.5 are frequently used (Hames 1981).

2.7. The Role of Electrical Parameters

Charged molecules are separated in an electrical gradient produced by a direct current (DC) as opposed to an alternating current (AC). Electrophoresis operates on two fundamental and interrelated electrical principles: (*a*) *electrical current* (I, amps), which is directly proportional to voltage (V, volts) and inversely proportional to resistance (R, ohms); mathematically, $I = V/R$; (*b*) *power* (P, Watts), which is directly proportional to the voltage and current. During the course of electrophoresis, heat is generated in the gel as a result of resistance due to the chemical and physical properties of the separation matrix. Mathematically, $P = V \times I$, and by substitution $P = I^2R$ (since $I = V/R$).

Buffer systems are designed to run at specified I, V, and P. In practice, one of these electrical parameters is held constant during electrophoresis. Power supply units are designed to deliver electricity at either constant power, voltage, or current. The properties of a gel and the electrode buffer change during electrophoresis. An effect of such electrophoretic changes could be a change in the resistance of the gel and consequently other electrical parameters. When the resistance of a gel changes, the electrical parameter that the researcher opted to hold constant (depending on the type of power supply unit) remains so while the other parameters change. Such changes in electrical parameters affect the mobility of the migrating protein molecules. For example, should resistance increase during electrophoresis while the voltage is held constant, it follows from the first electrical principle that the I (current, amps) will have to decrease accordingly. The consequence of a decrease in current is that the mobility of charged molecules will be slowed down but without any increase in heating of the gel, since I^2 will be small and, as Brewer and Sing (1970) pointed out, the heating of a gel is due more to I than V. Buffer systems are designed to run at high voltage and low current. On the other hand, if I (amps) is held constant in the above electrophoretic scenario, the mobility of charged molecules will be unaffected, since the rate of movement is directly proportional to I. The consequence of increased resistance at constant current is increased heating of the gel.

The ratio of the velocity of a particle to the electrical potential or field under which the particle is migrating is called the *electrophoretic mobility* (Zubay 1983). Mathematically, $\mu = V/E = z/f$, where V = cm/sec, E = electrical potential (volts/cm), z = net charge on particles, and f = frictional coefficient. The f value is larger for larger protein molecules, but for molecules of the same molecular weight, it is dependent on the shape of the molecules, i.e. it is larger for elongated, rod-like molecules than for globular ones. Furthermore, the force of electrophoresis is *E*, so the rate of migration is dependent on the net charge on the molecule rather than its mass (Zubay 1983).

2.8. The Effect of Molecular Size and Shape

As mentioned above, a separation medium may have sieving properties that influence the migration rate of charged molecules by retarding the mobility of large molecules while allowing small ones of identical charge to move with relative ease. The concept of *charge density* (charge per unit mass or length) may be introduced at this juncture. If molecules of different sizes and charges are being electrophoresed in a support medium with sieving properties, those with large charge densities will migrate faster than those with smaller charge densities. Certain electrophoretic procedures (SDS-PAGE) are able to cancel out the effect of charge so that molecules are separated on the basis of their molecular weight.

2.9. The Role of Temperature

The statement "when in doubt, keep it on ice" should be a rule of thumb in electrophoresis. The need to keep samples and reagents cold is emphasized repeatedly in this book. The heat generated by the process of electrophoresis must be dissipated through the provision of a cold environment. Excessive heat decreases enzyme activity. Cooling must be effected uniformly throughout the gel matrix. Uniform cooling is especially important when electrophoresing a thick gel (for many slices). The middle portion of a thick gel may not be cooled as well as the top and bottom parts. Although it is tempting to assume that the middle slices of a gel will produce the best-resolved bands, with improper cooling very thermolabile enzymes resolve poorly on middle slices. When supplemental cooling aids (water bag, ice pack) are used, it is critical that they cover the entire surface of the gel as uniformly as possible. Failure to do this may result in differential cooling and an irregular (curved) gel front.

2.10. Duration of Electrophoresis

Certain enzymes lose their activity rather quickly and are unable to survive prolonged periods of electrophoresis. These enzymes should not be included in runs that the researcher intends deliberately to prolong to achieve certain purposes. Generally, gels are run long enough to achieve good separation between bands. Since some enzymes migrate faster than others, the researcher must compromise in setting the duration of electrophoresis so that faster moving bands do not migrate off the gel while the researcher waits for slower bands to become clearly separated. With practice, one will be able to categorize the enzymes used as fast- and slow-migrating ones. If possible, one gel should be run and stained for enzymes in only one category so that slower enzymes may be run for a longer time without jeopardizing the faster ones. The duration of electrophoresis also will depend on the thickness of the gel being run. Thinner gels require a shorter duration of electrophoresis than thicker ones.

2.11. Protein Concentration of Samples

When homogenizing specimen samples for electrophoretic procedures that require samples to be absorbed into wicks for loading into gels, the tissue:extraction buffer ratio should be such that the extract has a consistency of a slurry. If need be, this slurry may be centrifuged to obtain a supernatant. If an extract is too dilute, there may not be enough enzyme to catalyze a specific reaction to produce a stainable product. In such a situation, the absence of a band could be misinterpreted as a null effect (lack of activity). Before declaring a null effect, it is advisable to repeat an assay. It should also be noted that the concentration of an enzyme is proportional to its catalytic activity (Mahler and Cordes 1968).

In polyacrylamide gel electrophoresis (PAGE) in which samples are loaded directly into wells, the quantity and concentration of the sample affects the results of electrophoresis. PAGE allows certain techniques to be employed (stacking) so that large amounts of dilute samples can be loaded to give satisfactory results.

2.12. Quality of Samples

"Garbage in, garbage out!" Samples for electrophoresis should, preferably, be fresh or else stored properly. Improper handling of samples during the collection stage can lead to reduced activity or even complete inactivity of enzymes. It is desirable to minimize the time that transpires between sampling and electrophoresis to ensure the superior quality of samples.

2.13. Sample Size

Researchers always have to decide on the sample size—number of single seeds, single plants, single animals, etc.—that will be statistically adequate for a purpose. Opinions on what is an adequate sample size vary, and it appears this issue must be decided by individual researchers according to the goal of the research. One of the applications of isozymes that requires a proper estimate of sample size is in the testing of seed purity in the seed industry. Nijenhuis (1971) first proposed electrophoresis as a tool for testing F_1 hybrid seed purity. The genetic model that formed the basis for his proposal was that each of the parents involved in a cross is fixed for a different allele at an isozyme locus. Therefore, only the authentic hybrid will display the heterozygous isozyme phenotype.

The assumption underlying the Nijenhuis (1971) genetic model is an optimal one and is not universally valid when isozymes are applied to authenticate hybridicity (Samaniengo and Arus 1983). Tanksley and Jones (1981) observed that very often the same allele is fixed in both inbred parents at all the isozyme loci analyzed, thus making estimation of the level of contamination by the Nijenhuis method impossible. Similarly, the fixed model is also invalid when one or both inbred parents are segregating at one or more loci. Samaniengo and Arus (1983) postulated a model that describes the data typically available for estimating contamination in hybrid seed production.

The literature indicates that a conservative number of 100 seeds may suffice for many purposes (involving diploid or tetraploid, self- or cross-fertilized species). Some studies, however, may require the researcher to use a statistical method to estimate the required sample size more precisely.

2.14. Quality of Reagents

Some reagents (especially stains) are light-sensitive whereas others need refrigeration or even freezer storage. Care must be taken to store chemicals properly to avoid deterioration. Recording the date of purchase on the container of each chemical is a good practice so that one can readily tell the age of supplies. Supplies with a short shelf-life should not be stocked in large quantities. Certain solutions (e.g. ammonium persulfate) for PAGE function best when prepared just before use.

2.15. Effect of Growth Environment and Developmental Stage on Isozymes

A classic example of the effects of the growth environment on enzyme activity is produced by alcohol dehydrogenase (ADH). A possible link between ADH polymorphism and soil moisture status has been suggested (Brown 1978). Assays of ADH can be improved when plants are stressed (by flooding) prior to sampling. This boost in activity is due to the role of ADH in anaerobic respiration (Johnson 1974). Gates and Boulter (1979) found seed tissue to be less sensitive to environmental fluctuation than other plant parts. A way of overcoming environmental influence is to generate all experimental materials under a controlled environment. Some enzymes show differential isozyme patterns at various stages of development (Scandalios 1969; Thurman et al. 1965), an event attributable to differential gene regulation. Age-dependent differences in patterns between young and older leaves for malic enzyme were observed in sugar beet (G. Acquaah, unpublished paper).

Results of electrophoretic studies indicate that an isozyme's pattern and intensity are specific to the plant part or tissue and to its maturity or developmental stage (Pierce and Brewbaker 1973). Eucaryotic enzyme loci are highly regulated and may be differentially repressed in different tissues and at different developmental stages (Gorman and Kiang 1978). In maize, catalase (CAT1) is active during the development of the kernel, but its activity decreases rapidly after seed germination (Scandalios 1979). Seasonal variation in the expression of esterase occurs in *Cucumis myriocarpus* (Schwartz et al. 1964). Variation in the growth environment, such as caused by disease, temperature, and nutrition, may affect the intensity of isozyme activity and sometimes produce novel bands (Hare 1966; Markert 1968). Some enzymes give consistent results, appearing to be unaffected by the age of the tissue provided that a growing leaf was sampled (Nielsen 1985).

2.16. Cryptic Variation

The properties of proteins may change when the constituent amino acids change as a result of mutations. Some changes go undetected because of their nature. For example, a change in a triplet code from TTT to TTC has no effect because both code for the same amino acid—lysine (redundancy of the genetic code). Likewise, a change from TCC (aspartic acid) to TCG (glutamic acid) does not change the net charge of the protein, since both are negatively charged. Sometimes, as a result of folding of proteins into more complex forms, some of the R-side chains are rendered inaccessible to ionizing agents and therefore do not contribute to the net charge of the protein molecule. The consequence of cryptic variation is that polymorphism is underestimated because the different forms of proteins cannot be effectively separated. By the same token, the polymorphism in a population can be overestimated when molecules become post-translationally modified by, for example, aggregation, such that multiple forms may be observed in electrophoresis even though enzymes are actually monomorphic ("pseudo-polymorphism").

2.17. Principles of Histochemical Staining of Gels

Enzymes catalyze the metabolism of substrates. Some enzymes are highly specific in their action whereas others have a wide range of physiological activity. During electrophoresis, enzyme molecules migrate to different locations in the gel. To mark the locations of these molecules, the appropriate substrate must be supplied for the enzymes to act on to produce a stainable product. The product of the action of an enzyme on a substrate must be visible so that it can be scored, but such products are not visible alone. Therefore, a chromatic effect must be induced by adding specific stains to the reaction solution that will interact with the product of enzyme activity to render it visible. Several enzyme-staining systems are in use. A detailed description of each system has been presented by Vallejos (1983). A brief summary of the rationale of the various systems is as follows.

2.17.1. The Azo Coupling System

The azo coupling system (ACS) has two main components, a diazonium salt and an aryl amine or alcohol (ester, amide). The basic reaction is

$$\text{diazonium salt} + \text{aryl alcohol} \xrightarrow{\text{coupling reaction}} \text{azo dye (chrommatic effect).}$$

ACS is used for staining hydrolytic enzymes (e.g. acid phosphatase, esterases, sulfatases, aminopeptidases, and glycosidases).

2.17.2. The Tetrazolium System

Most staining procedures use the tetrazolium system (TS). TS involves the use of tetrazolium salts (good electron acceptors), which are produced from the oxidation of formazans. During staining, the tetrazolium salt (soluble) is reduced to a formazan (colored precipitate). The commonly used tetrazolium salts are thiazolyl blue tetrazolium (MTT) and nitroblue tetrazolium (NBT): the former is more readily reduced than the latter (Pearse 1972). The electrons from a reducing substrate are accepted by a coenzyme (e.g. NAD^+ or $NADP^+$) and then transferred to an intermediary electron acceptor such as phenazine methasulphate (PMS). This acquired electron is later used to reduce a tetrazolium to a formazan. The enzymes in the oxidoreductase group use this system (e.g. malic enzyme, aconitate hydratase, alcohol dehydrogenase).

2.17.3. The Starch-Iodine System

Though a positive staining is possible, the starch-iodine system (SIS) is used primarily in negative staining protocols, which produce achromatic (white) spots. Enzymes that can be stained by SIS include amylase, phosphorylases, and catalase.

2.17.4. The Redox Dyes System

The redox dyes system (RDS) uses dyes whose physical and chemical properties change when their oxidation state changes, thereby making them become soluble or change color. An example of a redox dye is 2,6-dichlorophenol indophenol (DCPIP).

2.17.5. The Fluorescent Compounds System

The protocols that employ the fluorescent compounds system (FCS) produce molecules with the capacity to fluoresce (chromophores) when exposed to light of a certain wavelength. The general histochemical reaction involved in staining may be presented as

$$\text{substrate} \xrightarrow{\text{enzyme}} \text{primary product} + \text{dye} \longrightarrow \text{secondary product}$$

(the secondary product is a chromatic precipitate).

Take, for example, the staining procedure for shikimate dehydrogenase (SKDH), which employs TS. A staining recipe for this enzyme is

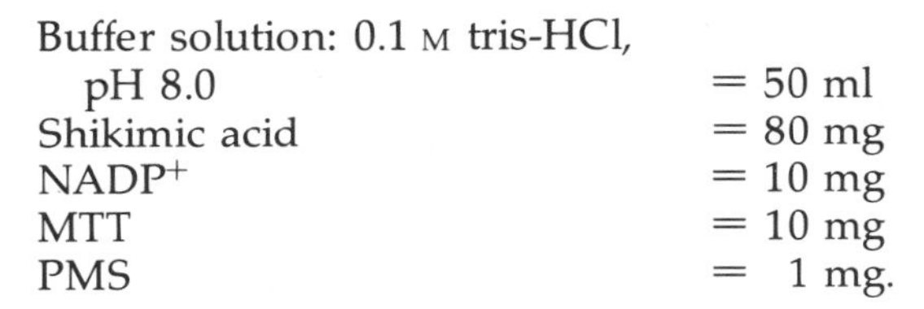

Buffer solution: 0.1 M tris-HCl, pH 8.0	= 50 ml
Shikimic acid	= 80 mg
$NADP^+$	= 10 mg
MTT	= 10 mg
PMS	= 1 mg.

The reaction involved in the staining process may be presented as shown in Figure 2-2.

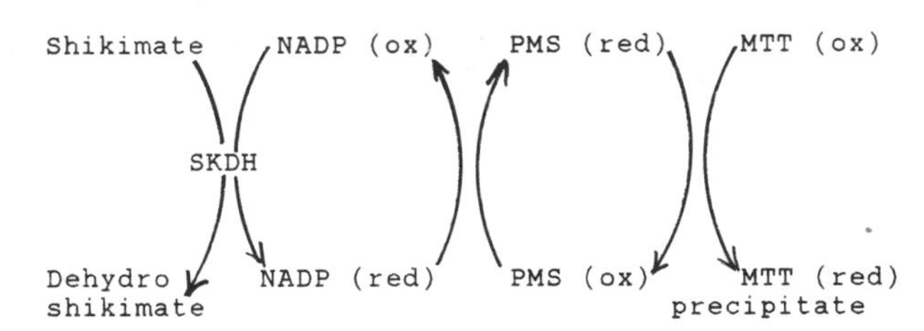

Figure 2-2. An example of the reactions involved in staining for an enzyme via the tetrazolium staining system.

In effect, a stain reveals the relative locations of the polymorphs of an enzyme that share a common catalytic activity. Unfortunately, suitable dyes have not been found for every enzyme. To circumvent this bottleneck, the reaction of one enzyme may be coupled to another enzyme that is downstream in a metabolic pathway by directly adding all the intervening enzymes to the stain solution. The product of the action of one enzyme becomes the substrate for another. The staining protocol for phosphoglucose isomerase (PGI) is a good example. A recipe for PGI is

Stain buffer: 0.1 M tris, pH 8.0	= 50 ml
Fructose-6-phosphate	= 80 mg

$MgCl_2$	= 40 mg
$NADP^+$	= 5 mg
G-6-PDH	= 2.5 l
MTT	= 5 mg
PMS	= 5 mg.

The intervening enzyme supplied in the above staining solution is glucose-6-phosphate dehydrogenase (G-6-PDH). The reactions for this metabolic process are

$$\text{F-6-P} \xrightarrow{\text{PGI}} \text{G-6-P}$$

$$\text{G-6-P} + \text{NADP}^+ \xrightarrow{\text{G-6-PDH}} \text{6-PG} + \text{NADPH}.$$

The product of the first reaction, G-6-P, is used as a substrate for the second reaction. The above recipe for PGI shows two other components, which are not required in all staining solution recipes. The first is the coenzyme ($NADP^+$ in the present example). Amino acids lack certain side chain functional groups that enable them to catalyze certain reactions. To make up for the deficiency, they rely on "helper" molecules called *coenzymes* that possess the requisite properties for the reaction (Zubay 1983). Whenever $NADP^+$ is required in a recipe, the researcher should use the β-form (i.e. β-$NADP^+$). Furthermore, when a coenzyme is required in a reaction, the electrophoretic mobility of proteins may be affected by the degree to which the enzyme is saturated with the appropriate coenzyme (Micales et al. 1986). It follows, therefore, that although coenzymes (especially $NADP^+$) are expensive and must be used sparingly, the researcher must be careful not only to supply at least the minimum amount prescribed for the reaction, but also to ensure that equal amounts are supplied for all the reactions in a particular study. This will forestall any artifactual effects from erratic mobilities induced by varying amounts of a coenzyme.

In addition to a coenzyme, some enzyme reactions require the presence of organic or inorganic molecules ($MgCl_2$ in the present example). These kinds of helper molecules are called *cofactors* (Zubay 1983).

Although in many cases isozymes are identified by colored bands (positive staining) (Figure 2-3), the dye-substrate interaction sometimes produces a clear spot (negative staining), as is the case in catalase (Figure 2-4).

When coupling of reactions is needed and hence several intermediate products are present, some of which may be soluble, diffusion of soluble products through the gel may sometimes occur and blur the otherwise well-defined bands. To reduce the undesirable blurring of bands, Markert and Moller (1959) developed the agar overlay technique whereby the reagents in a stain recipe are mixed with agar and applied to the gel. As the agar cools, it forms a solid transparent layer through which bands are visible. A modification of the agar overlay technique is the filter paper overlay technique developed by Scopes (1968). The agar in the agar overlay technique is replaced by a piece of paper (filter paper such as Whatman Nos. 1, 2, or 3 or cellulose acetate paper). Overlays also are used when the enzyme concentration is low or its activity is weak.

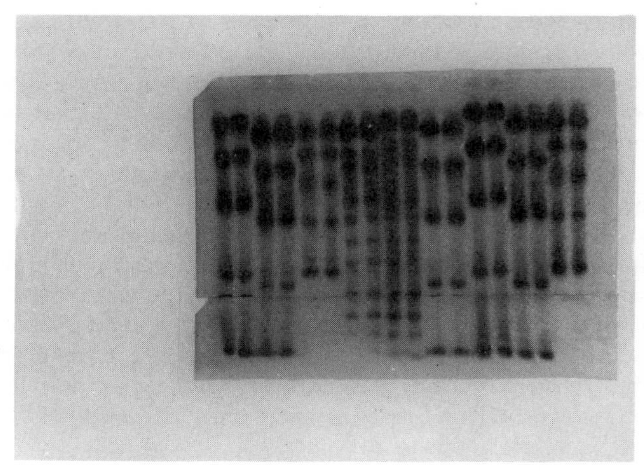

Figure 2-3. A zymogram of a positively staining enzyme. (Photograph provided by Mary Malmberg.)

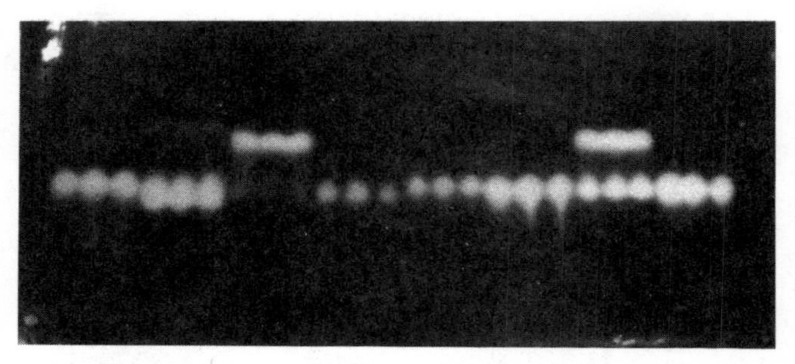

Figure 2-4. A zymogram of a negatively staining enzyme (catalase of *Phaeoisariopsis griseola*, the causal fungus of angular leaf spot of dry bean). (Photograph provided by Lucia Afanador.)

Some reagents of a staining solution are very expensive, and hence must be used economically. In such cases, instead of preparing the usual 50 ml (or 100 ml) solution, the recipe may be proportionally reduced to make about 6–10 ml and applied to one surface of the gel by using a Pasteur pipette. This "microstaining" technique has the added advantage of reducing diffusion of soluble products through the gel.

Kinetics are another important consideration in enzyme-catalyzed reactions. One of the kinetic parameters commonly encountered is the Michaelis constant (K_M) which is a function of the substrate, temperature, pH, and ionic strength (Zubay 1983). As catalysts, enzymes speed up the rate of biochemical reactions. The time required for staining is dependent on the K_M of the enzyme. The researcher sooner or later will find that certain stains develop almost instantly (e.g. staining for catalase) whereas others require several hours. K_M values for some of the enzymes featured in this book are catalase (substrate is H_2O_2) = 1.1, aspartate aminotransferase (substrate is α-ketoglutarate) = 1.0×10^{-4}, and fumarase (substrate is fumarate) = 5.0×10^{-6}.

When purchasing an enzyme, notice that the label on the container indicates the enzyme unit. This is the amount of substrate that is converted per unit time by the enzyme under specified conditions of pH, substrate concentration, temperature, etc. (Mahler and Cordes 1968).

2.18. Staining Solution Buffers

Substrates, intermediate enzymes, and activity stains are delivered to a gel in a well-buffered solution suitable for the metabolic reactions involved. As expected, the pH is a critical factor in the staining process. Staining protocols have recommended pH values that are not universally applicable. Significant improvements in resolution of isozymes when pH was manipulated have been recorded. Soltis et al. (1983) modified the pH recommendations of 7.0 or 7.1 for ALD, GDH, IDH, MDH, ME, PGI, PGM, and 6-PGD reported by Shaw and Prasad (1970) to between 8.0 and 8.5 for best results in ferns. O'Malley et al. (1980) observed that the composition of a stain solution was not critical as long as its buffering capacity was adequate for the metabolic reaction in question.

Tris buffers are very versatile and are widely used at various strengths and pHs for preparing electrophoretic buffer solutions; however, non-tris buffers give better results in certain cases. A few of the numerous staining solution buffers are presented in Table 2-2 to show the range of pH and examples of enzymes that have been successfully assayed with such buffers. The recipes given are for preparing one liter volume of solution.

Table 2-2. Selected staining solution buffers.

Buffer	pH	Recipe	Enzyme
0.1 M tris-HCl		tris = 12.125 g H_2O = up to 1 l	
	9.1		IDH, SOD, MDH[a]
	7.0–7.1		ALD, GDH, G6PDH, HK, MDH, ME, PGI, PGM, 6PGD[b]
	8.0–8.5		Same as for 7.0[c]
0.05 M tris-HCl		tris = 6.062 g H_2O = up to 1 l	
	8.0		IDH, MDH, PEP, 6PGD, PGI, PGM[d]
0.2 M tris		tris = 24.25 g H_2O = up to 1 l	
	8.0		ACO MPI, DIA[a]
0.1 M tris-maleic		tris = 12.1 g Maleic acid = 11.6 g NaOH = 1.6 g H_2O = 1 l	
	5.5		ENP[a]
0.1 M Na. Acetate		Na. Acetate = 13.6 g 1 N HCl = 24.25 ml H_2O = up to 1 l	
	5.0		ACP[a]

[a]Cardy and Beversdorf (1984). [b]Shaw and Prasad (1970). [c]Soltis et al (1983). [d]Conkle et al (1971).

2.19. Strategies for Quick Staining

Preparing staining solutions can be time consuming. To quicken the staining process, the required amount of some items may be pre-weighed into suitable containers (such as microcentrifuge tubes) and stored appropriately until needed. Weighing every reagent accurately may not be necessary. With experience, one may use a spatula to estimate some of the chemicals and thereby save time. Expensive chemicals, however, should be accurately weighed. Items such as stains and some substrates and enzymes may be put into solution ahead of time and stored in the required amounts or pipetted out when needed. Since many of the stains are light-sensitive, their solutions must be stored in light-proof bottles. Freshly prepared staining solutions are preferable but not always convenient, so stock solutions may be prepared. Such stock solutions should be stored in a refrigerator, where they stay in good condition for several weeks. Stain buffers may also be prepared ahead of time and stored in a refrigerator. When using any chemical that may be unstable in solution, prepare a fresh solution for each use or prepare amounts to last only a very short time.

Some of the chemicals that may be prepared ahead of time and kept in solution are listed in Table 2-3. Working with solutions may be time-saving, especially when one has a heavy schedule of work. Furthermore, weighing 100 mg accurately and quickly is easier than weighing 10 mg. Stock solutions should be prepared in concentrations that can be conveniently delivered and batched in amounts used frequently in recipes. For example, 10 mg of $NADP^+$ or multiples of it is the usual amount required in staining solution recipes, so a convenient concentration of $NADP^+$ stock solution to prepare is 10 mg/ml. Those who prefer to use only fresh ingredients but like the convenience of solutions may prepare a bulk amount of solution sufficient for a specific amount of work.

Researchers who prefer to use dry compounds may quicken the staining process by preweighing some of the chemicals and storing them until they are needed

Table 2-3. Guide to preparing stock solutions of some staining solution chemicals.

Chemical	Suggested concentration	Required for 25 ml[a]	Storage type
Stains			
Fast blue BB	50	1.25 g	F, LP
MTT	10	250 mg	R, LP
NBT	10	250 mg	F, LP
PMS	5	125 mg	F, LP
Substrates			
Aconitic acid (pH 8)	100	2.5 g	R
Malic acid	100	2.5 g	R
Naphthyl acid phosphatase	100	2.5 g	R
Fructose-6-phosphate	50	1.25 g	R
6-Phosphogluconic acid	20	500 mg	R
Enzymes[b]			
Glucose-6-phosphate dehydrogenase	50 units (1000 μ/ 20 ml)		R
Phosphohexose isomerase	100 units (1000 μ/ 10ml)		R
Coenzymes			
NADP	10	250 mg	R
NAD	20	500 mg	R
Cofactors			
Magnesium chloride	100	2.5 g	R

[a]R, refrigerator; F, freezer; LP, lightproof container.
[b]Enzymes may be kept in microcentrifuge tubes.

Table 2-4. Some preweighed (dry) staining solution chemicals.

Chemical	Suggested amount for 50 ml solution	Storage type[a]
Isocitric dehydrogenase	20 mg	F
Mannose-6-phosphate	20 mg	F
6-Phosphogluconic acid	20 mg	F
Glucose-1-phosphate	250 mg	F
Diazo blue B	150 mg	R
PVP-40	300 mg	RT
Glutamic acid	150 mg	RT
Calcium chloride	50 mg	RT

[a] F, freezer; R, refrigerator; RT, room temperature.

(Table 2-4). These preweighed chemicals may be put into microcentrifuge tubes (1.5 ml, 0.5 ml). The lists of chemicals in Tables 2-3 and 2-4 are only examples. The researcher may modify the amounts as desired and may add to the lists. Stains tend to leave marks in the staining dishes and beakers and on other apparatus such as magnetic stirring bars and beakers. Labeling and restricting the use of such apparatus to specified enzyme stains is recommended. This may help reduce the incidence of stain artifacts. Thorough washing after staining is especially required if dishes are not restricted to staining specific enzymes. Preparing staining solutions in graduated beakers eliminates the use of measuring cylinders to measure out each stain buffer.

After assaying different enzymes over a period, one may discover that some isozyme bands are always located closer to the anodal end of the gel (i.e. they have fast mobility) whereas others hardly migrate at all (i.e. they have slow mobility). It is possible to split a gel into two (or more) pieces and stain each one for a different enzyme! This has been tried successfully with phosphoglucose isomerase (PGI) (fast) and fluorescent esterase (FLE) (slow) in soybean (C. Sneller, personal communication). O'Malley et al. (1980) suggested that the slice of gel used for staining for FLE could even be used for another enzyme after scoring. Furthermore, *combination staining,* whereby two enzymes are simultaneously stained for on the same slice of gel, is also possible. The appropriate substrates and other chemicals required are supplied in the same staining solution for combination staining. In common bean, superoxide dismutase (SOD) may be scored from a gel stained for alcohol dehydrogenase (ADH) or shikimate dehydrogenase (SKDH). In fact, SOD appears voluntarily on many gels stained via the tetrazolium system with MTT and PMS. Stuber et al. (1988) stained for PGI and PGD on one gel. Although a combination staining strategy can be developed by trial and error, some guidelines are as follows:

1. Be familiar with the patterns of the two enzymes in order to distinguish between the sets of bands they produce.
2. The two sets of bands must not overlap.
3. Scoring a gel is easier if the two enzymes involved stain with different colors.
4. The two stains must not interact.
5. Enzymes with similar staining schedules offer an opportunity for combination staining (e.g. MDH and ME, PGM and PGI).

The following are some suggestions to help make staining of gels quicker and easier. Prepare recipe cards that list the chemicals and their required quantities in the order in which they are added to the staining buffer. Some solid ingredients may have to be added to the buffer ahead of time and stirred to dissolve. If several enzymes are being assayed at the same time, color code the recipe cards and the buffer containers so that when a buffer container with, for example, a blue label is picked up, one knows immediately that it is required by all cards with blue labels (Figure 2-5).

Solid substrates are usually added and stirred into solution before other staining solution ingredients are added. Logically, enzymes are added before stains, but the complete solution should not be prepared more than a few minutes ahead of time because formazan may start forming prematurely in some staining solutions. For best results, move quickly from one step in the staining process to the next.

Enzyme: PGM

	Amount	Check
Buffer: 0.1 M tris-HCl pH = 8.0	50 ml	
Fructose-6-phosphate	80 mg	
$MgCl_2$	40 mg	
G-6-PDH	2.5 μl	
NADP	5 mg	
MTT	5 mg	
PMS	0.5 mg	

Instructions:
Incubate at 37° for 30–60 min.
Do not overstain; rinse and fix.

Figure 2-5. A sample staining recipe card.

Set out the staining dishes ahead of time in the order of staining (Figure 2-6). This step is important because some enzymes do not resolve clearly on gel slices taken from certain parts of the stack (top, bottom, or middle). Improper cooling may cause certain parts of the gel to be warmer than others so that enzymes very sensitive to heat are adversely affected. The topmost slice is usually discarded. During a lengthy staining process, all stains and other required reagents should be placed in a tray of ice to keep them cold. The gel should be sliced before the preparation of the stain solution is completed, since some solutions start to deteriorate shortly after preparation. The best results are obtained when certain staining solution components are added just before the solution is poured onto a gel (check protocols). Certain staining procedures may require at least partial darkness, but most procedures can be performed under ordinary laboratory conditions. Dimming the lights in the laboratory during staining for the sake of the light-sensitive components of the stain solution may be helpful in preventing deterioration of the stain compounds. Some of the dark background of stained gels is attributable to excessive exposure of stain solutions to light (Vallejos 1983). Amber-colored or stained glass flasks are available if one prefers to use them.

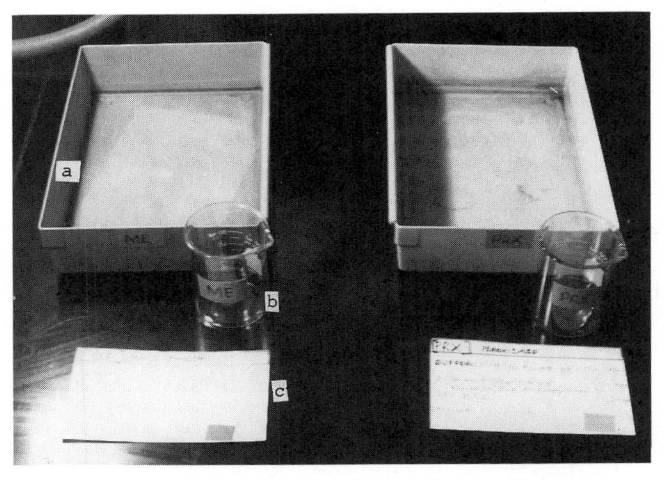

Figure 2-6. Setting up apparatus for staining a gel: (*a*) staining tray; (*b*) staining solution container (e.g. glass beaker); (*c*) recipe card. (An ice tray should be provided for cooling chemicals that are removed from storage in the freezer/refrigerator for preparing the staining solution.)

To begin staining, pour the complete staining solution into the staining dish or tray first (gels tend to stick to the bottom of dishes when there is no solution). Then take the dish to the place where the gel was sliced (or take the slicing bed with gel slices to the dish to minimize the chances of tearing the gel) and carefully peel off a slice, holding it in both hands and with as many fingers as possible. Transferring a slice of gel to a staining dish is a delicate operation, especially when handling gels of low starch concentration that produce very fragile slices. Gently set the slice of gel in the staining solution and swirl to ensure that the gel is completely covered with stain. The slice of gel must be submerged in the solution for proper and uniform staining, so choose a dish of a size that will ensure submergence of the gel at the chosen volume of staining solution. If a gel gets torn in the process of handling, place all the pieces in the staining tray and piece together after staining. Cover the dish and place it under the prescribed incubation conditions. An aluminum foil cover helps to create dark conditions in the dish.

Most procedures require incubation at room temperature (27°C +/− 3) but a higher temperature of 37°C is frequently recommended for quicker results. If gels do not have to be incubated, the trays must be covered with light-proof material or placed in a dark place (unless otherwise indicated). Different enzymes have different lengths of time of incubation. Some enzymes "develop" (chromatic bands appear) within 30 to 60 minutes; others require several hours. When some enzymes are left in the staining solution for too long they become "overdeveloped" (too darkly stained) and may be unscorable. For certain enzymes, bands appear to develop in stages. A set of bands appears relatively early and is usually more intensely stained (primary bands). If staining is prolonged, other bands appear (secondary bands). Keeping track of the patterns of the development of bands during staining is recommended. Information on band development is helpful when banding patterns are being interpreted and also for model building. Certain buffer systems eliminate the secondary bands, thus making interpretation of zymograms sometimes less complicated. In addition to secondary bands, primary bands may sometimes be accompanied by "shadow bands," which in some cases can be removed by changing electrophoretic conditions (e.g. addition of EDTA).

Since it is very possible to forget to add an ingredient or two to the staining solution during preparation (especially when handling several enzymes at the same time), checking the gels within the first 15 to 30 min of the incubation process is recommended. If one does not see familiar signs, an ingredient may have been omitted from the solution. Salvaging a gel is sometimes possible by adding the item(s) suspected to be missing from the staining solution. To reduce the chance of omitting a stain component, it is good practice to laminate the recipe cards in plastic, or alternatively, to fix a transparent tape over the section indicating the amounts of chemicals required (Figure 2-5). With a water-soluble marker, check off each item as it is added to the solution. After each staining session the marks should be removed.

It is best to stain soon after electrophoresis is completed to prevent unnecessary diffusion of proteins in the gels, which causes poor resolution of bands. If immediate staining is not possible, the gel may be wrapped in a plastic wrapper and stored in a refrigerator for no more than about two hours to avoid diffusion of enzyme bands in the gel.

When the tetrazolium system is used for staining, some gels show continuously stained streaks starting from the origin to the first distinct band. This is attributable to the reducing agents (e.g. 2-mercaptoethanol, reduced glutathione) included in some grinding buffers (Vallejos 1983). The reducing agents move through the gel and reduce tetrazolium salts at the migration front.

REMINDER

1. Protect light-sensitive compounds from direct light.
2. If staining buffer solutions are stored in a refrigerator, the required amounts should be poured out into staining containers prior to the preparation of the staining solution to allow some warming to occur. Remember, the incubator temperature most recommended is 37°C.
3. Wear gloves (and a face mask sometimes) when handling the stains.
4. Do not pour stain solution (fresh or spent) down the drain! Dispose of waste chemicals according to the prescribed procedures for hazardous wastes.
5. Overstaining could be most undesirable for some enzymes.
6. Keep track of the trends in the development of bands.

2.20. Terminating the Staining Process

As soon as the bands have appeared to the desired intensity, the staining process should be discontinued. Score the gel immediately after staining is discontinued, since certain enzyme bands fade away rather quickly and some enzymes are sensitive to overstaining. Furthermore, for certain stains, a prolonged staining period causes a film of stain-starch complex to form on the gel. This film may be removed by squirting water from a wash bottle onto the gel so the bands become more visible. To terminate the staining process, first drain out the staining solution and rinse the gel with tap water. When rinsing gels, the process is made easier if a piece of flexible plastic gauze is placed on the gel in the tray. Do not press too hard on the gauze. After rinsing, the gel should be scored without delay and fixed or photographed.

2.21. Limitations of Electrophoresis

Routine electrophoresis detects only the amino acid substitutions that result in differences in the net charge of proteins (Ayala and Kiger 1980). It is this difference in the net charge that causes the enzymes to have different electrophoretic mobilities (i.e. move at different rates in an electric field). A single amino acid substitution (a base pair substitution of the DNA, a point mutation in a structural gene) can change the electrophoretic mobility of the protein (Ingram 1957).

As sensitive as electrophoresis may be, only about a third of all amino acid substitutions can be detected by the technique (Lewontin 1974). Consequently, the amount of polymorphism or allelic variation will probably be underestimated by electrophoresis as a result of this cryptic variation.

Electrophoresis is also limited by the number of available staining protocols. Of the available ones, not all can be successfully employed in every situation. Staining protocols for over 50 enzymes are provided in this book. The protocols available are for water-soluble proteins.

Another restriction on the utility of electrophoresis in the detection of genetic variation is that only the variability in the coding portions of the DNA (which constitute just about 10% of the total eukaryotic genome) can be sampled. In other words, only transcription and translation products of structural genes are accessible to electrophoretic analysis.

2.22. Tips on Preparing Solutions

2.22.1. Caution in the Laboratory

Certain stains are hazardous to health (e.g. MTT). Be sure to know which stains are potentially harmful and handle them with care (wear protective clothing). Avoid touching staining solutions and wash frequently if handling chemicals with bare hands.

2.22.2. Concentration of Solutions

1 molar (M) = formula weight of solute/liter of water; 1 normal (N) = gram equivalent weight (g.e.w.) of solute/liter of water. Therefore, M × (no. g.e.w./mol. wt.) = N. Gram equivalent weights of common acids and bases are

HCl = 36.46 g	$NaOH$ = 40.00 g
H_2SO_4 = 49.04 g	KOH = 56.11 g
$Ca(OH)_2$ = 37.05 g	

2.22.3. Stock Solutions

Prepare stocks of common buffer solutions (e.g. 1 M tris) and store to dilute and titrate to the appropriate pH as required. Refrigerate all stock solutions. Buffer solutions may be prepared in stronger than needed concentrations (e.g. 5×, 10×). This allows the solutions to be stored in smaller containers, thereby saving shelf space. The concentrated stock solutions are diluted with water to the appropriate concentrations when needed.

2.22.4. Diluting Concentrated Stock Solutions

A general formula for conversion to obtain a specified amount of a dilute solution from a concentrated one is

$$C_R V_R = C_A V_A$$

where V_R is the volume of diluted solution required, and C_R and C_A are the concentrations required and available, respectively. V_A is the volume of concentrated solution to be diluted. For example: 100 ml of 0.05 M solution is required (to be prepared) from a 1 M solution. Using the formula,

$$C_R V_R = C_A V_A$$
$$V_R = C_A V_A / C_R$$
$$100 = 1 \times V_A / 0.05$$
$$V_A = 100 \times 0.05/1$$
$$= 5 \text{ ml.}$$

The amount of concentrated solution available is 5 ml; therefore to obtain 100 ml of diluted solution, the amount of water to add is 100 − 5 = 95 ml.

REMINDER

1. Whereas many chemicals used in buffer recipes may not be very hazardous to handle, some of them [e.g. *N*-(3-aminopropyl)-morpholine] must be handled with care. It is advisable to treat all chemicals as potentially hazardous (consult the Merck™ index).
2. Phosphate buffers are prone to bacterial growth when stored for prolonged periods without refrigeration.

Chapter 3

STARCH GEL ELECTROPHORESIS

The basic apparatus for starch gel electrophoresis (SGE) is described in this chapter. By following the list of materials provided and the descriptions of the basic models of SGE units, a researcher will be able to construct a homemade unit to the desired practical specifications. One may choose to construct electrophoresis units larger than the examples given in this chapter, but bear in mind that starch gels molded from units larger than the "regular size" molds may be more difficult to handle without tearing the gels.

3.1. Apparatus for Starch Gel Electrophoresis

The major hardware required for SGE that are unlikely to be already available in a modestly equipped biology laboratory are (*a*) source of DC power supply, (*b*) buffer/solution containers, (*c*) gel molds, (*d*) pH meter, (*e*) electrolyte tanks, (*f*) incubator/oven, (*g*) staining trays, and (*g*) freezer-refrigerator.

In addition to the above items, most of the chemicals needed for electrophoresis are unlikely to be in stock in a general laboratory. A comprehensive list of equipment and supplies for setting up a laboratory for electrophoresis is provided in Chapter 9.

Many researchers prefer to construct their own gel molds and electrolyte tanks to meet specific research needs and, more importantly, to save money. Ready-made gel apparatus can be extremely expensive. There are two basic models of SGE apparatus, and many modifications are possible to suit individual laboratories' needs. Each model consists of one gel mold and a pair of buffer tanks. The materials needed to construct either model are essentially the same, the minor differences lie in the amounts needed. The dimensions of a mold should be chosen with several considerations in mind (Brewer and Sing 1970). Some of these are as follows:

1. The mold should be long enough to permit adequate duration of electrophoresis to clearly separate the bands without jeopardizing the faster migrating isozymes (i.e. not so long that they migrate into the buffer tank).
2. The depth of a gel mold is dictated by the thickness of the gel to be molded, which in turn is based on the number of slices of gel needed. Casting a gel too thick, however, predisposes its interior to heating during electrophoresis. A depth of 1 cm is usually adequate for a gel mold for most purposes.
3. The slices obtained from gels of low starch concentration are very thin and delicate and require extra care in handling. The problem of handling is compounded when the gel mold is very large.

The following is a typical list of materials needed to construct a gel mold and two electrolyte tanks. Some modifications may introduce new components not featured in the list.

1. Clear acrylic, plexiglass, or perspex of different thickness: 0.15 cm for spacers for slicing, 0.3 cm for the gel mold or buffer tanks, 0.8 cm for the slicing bed and block, and 1.2 cm for the pestle.
2. Bandsaw or a small hacksaw for cutting the pieces of plexiglass.
3. Platinum wire: 24 gauge or stainless steel wire for use as electrode wires.
4. Sockets for electrodes: choose color-coded (black and red) type.
5. Methylene chloride for gluing the pieces of plexiglass together.

6. Soldering iron for connecting the electrode wires to the terminals.
7. Patch cords (color-coded) for connecting the buffer tanks to a power supply.

The thickness of plexiglass may be changed if so desired. What is essential is that the finished product be rigid enough for the purpose for which it was constructed.

3.2. Gel Mold and Buffer Tank Type I: Wick Model

The wick model is the older of the two gel mold models. In the Type I model, the buffer tanks are usually fixed to a bed to form a single piece (Figure 3-1), but several modifications also work well. In one such modification, the gel mold consists of a bed and loose wall pieces, which must be taped together before the starch gel is poured into the mold (Figure 3-2).

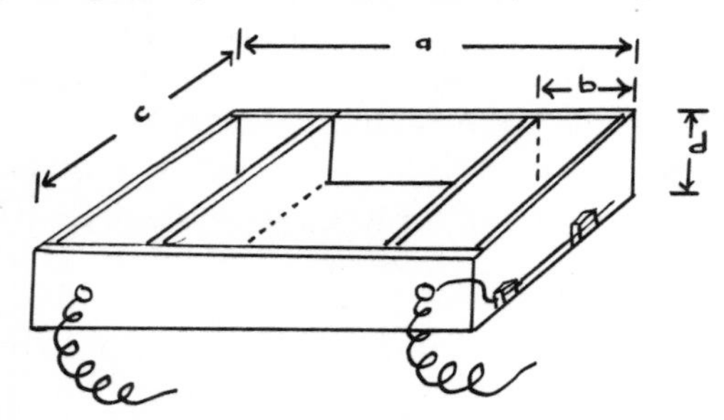

Figure 3-1. A Type I electrophoresis buffer tank: $a = 30$ cm, $b = 5$ cm, $c = 20$ cm, $d = 5$ cm. Each tank can hold up to 500 ml of buffer.

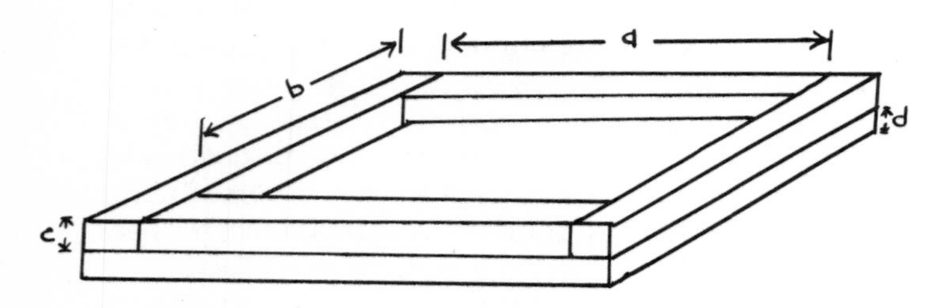

Figure 3-2. A Type I electrophoresis gel mold: $a = b = 17$ cm, $c = 10$ mm, $d = 3$ mm. The walls may be loose or fixed. This unit will hold about 330 ml of starch gel and provide up to 6 slices of gel.

Instead of taping, the walls may be held in place with paper clamps. The walls may also be fixed together and to the bed to form one rigid piece. To complete the electrical circuit for electrophoresis, the gel is connected to the electrolyte in the tanks by means of wicks, which may be made from paper (handywipes), sponge, cloth, or other suitable absorbent material (Figure 3-3). The wicks must be able to stay saturated with buffer solution throughout the electrophoresis or else voltage fluctuations will occur. The dimensions of the pieces were obtained from Weeden and Emmo (n.d.). By using 2 mm wide wicks, one can load about 20 samples at a good spacing in this size of gel. An additional ten samples may be loaded if one is quite familiar with the technique and the enzymes being studied. Crowding can sometimes complicate electrophoretic results. A spacing of 1 mm should be the minimum distance between adjacent wicks.

Figure 3-3. A Type I electrophoresis unit showing the position of the gel mold on the buffer tank.

3.3. Gel and Buffer Tank Type II: Wickless Model

The wickless mold was introduced by Johnson and Schaffer (1974). The major difference between the Type I and Type II models is the presence or absence of wicks. Also, the gel mold in the Type II model has permanently fixed parts and resembles a short table with two hollow "legs" (Figure 3-4). The buffer tanks may be detached (Figure 3-5) or fixed to a bed as in the Type I model. The opening in the lower part of each leg is taped shut *before* the gel is molded and removed to expose the gel (rectangular strip) *before* electrophoresis. Each leg sits in a buffer tank with the exposed gel in it making direct contact with the electrolyte solution, thus eliminating the need for a wick (Figure 3-6).

The Type I model is by far the more commonly used, but both models are used successfully in various laboratories. The choice of a model is, therefore, a matter of preference. Either model has advantages and disadvantages. The Type II model is more difficult to construct without professional help. Furthermore, the elimination of the wick in the Type II model increases by about ⅓ the starch needed to mold a gel. In the Type I model, the wick sometimes does not stay saturated during electrophoresis. Also the Type I model with the loose walls is more cumbersome to set up because it requires the pieces to be taped down to the bed each time a gel is to be cast. An advantage, however, is that the gel mold serves as a slicing bed after the walls are removed.

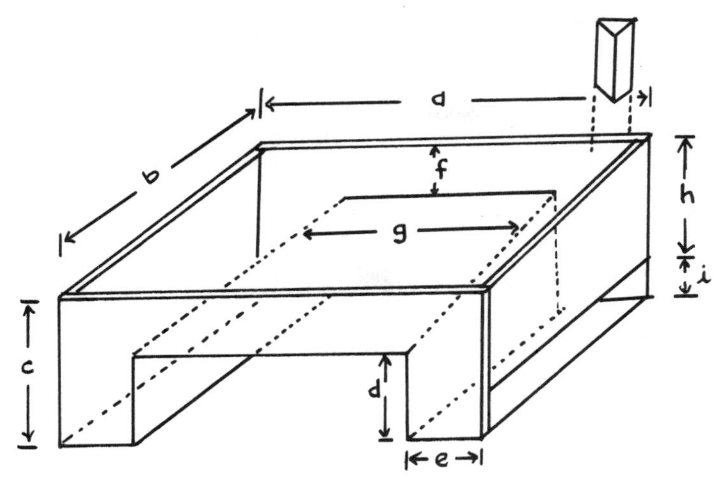

Figure 3-4. A Type II electrophoresis gel mold: $a = 19.2$ cm, $b = 18.3$ cm, $c = 3.9$ cm, $d = 2.6$ cm, $e = 1.6$ cm, $f = 1.3$ cm (thickness of gel), $g = 16.0$ cm, $h = 2.6$ cm, $i = 1.3$. This tray will contain up to 500 ml of starch gel for 6 slices of gel.

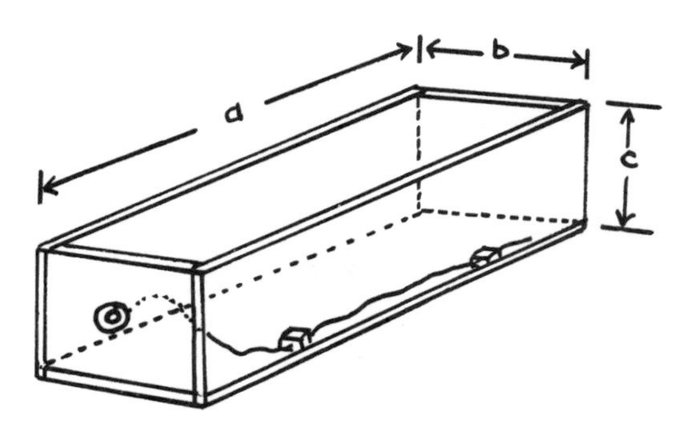

Figure 3-5. A Type II electrophoresis buffer tank: $a = 20$ cm, $b = 6.25$, $c = 5$ cm. This tank can hold up to 500 ml of buffer.

Figure 3-6. A Type II electrophoresis gel unit showing the position of the gel mold on the buffer tanks.

3.4. Selected Buffer Systems for Starch Gel Electrophoresis

As noted above, both types of buffer systems have their advantages and drawbacks. Discontinuous systems generally produce sharper band definitions than continuous ones. In SGE, this enhanced band definition results from the front of the electrolyte moving into and through the gel, forming an interface of the two unlike buffers. The interface appears to have a "compressing" effect on the migrating pack of protein molecules. For example, one may observe the effect of a discontinuous buffer on band definition on a lithium-borate/tris-citrate electrophoresis buffer system. The tracking dye starts off as a small blob within the initial 30 min or so. After about 2 hr of electrophoresis, the dye is compressed into a thin line at the migration front! In a continuous buffer system, however, the marker dye lane becomes progressively larger and streakier with time. Continuous buffer systems are easier to prepare because one simply dilutes the electrode buffer to a specified strength to obtain the gel buffer.

Some buffers are more versatile than others and are able to resolve many different kinds of enzymes. One buffer system is unlikely to be adequate for assaying many enzymes, however. Researchers therefore routinely use two or even three systems for both separation of proteins and confirmation of results. Different systems may be required to resolve completely all the regions of activity for some enzymes. Furthermore, different systems may reveal different patterns for the same enzyme. Polymorphism was detected at certain enzyme loci in beans only when specific buffer systems were used (Sprecher and Vallejos 1989). Certain buffers are able to eliminate minor or secondary bands, which often complicate the interpretation of electrophoretic results. Using one set of electrophoretic conditions may exclude the detection of certain subtle (cryptic) genetic variations of structural genes (Coyne 1982; Moore and Collins 1983). Shumaker et al. (1982) suggested varying gel and/ or electrode buffer pH and starch concentration in electrophoresis to help improve the quality of results, their interpretation, and reliability. In other words, employing several buffer systems in research is recommended.

The following list represents some of the common buffer systems in use. Modifications with respect to pH and molarity abound. A working recipe for each buffer system has been supplied with quantities of component reagents given on a per liter basis. Where available, the original electrical specifications (volts × amps) and duration of electrophoresis have been indicated. Where no duration of electrophoresis and power required have been indicated, the researcher should determine these parameters by trial and error. The duration of SGE is usually between 4 to 6 hr at power settings of between 15 and 30 Watts. The amount of current per gel is between 40 and 70 mA for regular runs. The power settings will vary depending on the concentration of the medium ingredient, gel thickness, ionic strength of the buffer, and the chosen speed of run (fast or slow). All solutions are prepared with distilled water. The selected buffer systems are as follows:

1. Morpholine-citrate system (Clayton and Tretiak 1972).

   (*a*) Electrode buffer (pH = 6.1)
   0.04 M Citric acid (anhydrous) = 7.68 g
   Water = 1 l
   [adjust pH with *N*-(3-aminopropyl)-morpholine] = 10 ml

   (*b*) Gel buffer (pH = 6.1) (1:20 dilution of electrode buffer)
   0.002 M Citric acid (anhydrous)

   POWER: 150 V × 50 mA

2. Lithium-borate/tris-citrate system (Ridgway et al. 1970).

   (*a*) Electrode buffer (pH = 8.1)
   0.06 M Lithium hydroxide = 2.52 g
   0.30 M Boric acid (or 0.15 M boric oxide) = 10.45 g
   Water = 1 l
   [adjust pH with dry components]

   (*b*) Gel buffer: tris-citrate (pH = 8.5)
   0.03 M tris = 3.63 g
   0.005 M Citric acid (anhydrous) = 0.96 g
   Water = 1 l

   POWER: 250 V (raise to 300 V after 1 hr)

3. Borate/tris-citrate system (Tanksley 1979).

   (*a*) Electrode buffer (pH = 7.8)
   0.3 M Sodium borate = 18.55 g

   (*b*) Gel buffer (pH = 7.8)
   0.0036 M Citric acid = 0.756 g
   0.00152 M tris = 0.184 g
   Water = 1 l

4. Borate/tris-citrate-borate system (Aston and Braden 1961).

   (*a*) Electrode buffer (pH = 8.3)
   0.19 M Boric acid = 11.74 g
   0.14 M NaOH = 5.60 g

   (*b*) Gel buffer (pH = 8.3)
   0.046 M tris = 5.57 g
   0.68 M Citric acid (anhydrous) = 130.60 g
   0.01 M LiOH = 0.42 g
   0.019 M Boric acid = 1.17 g
   Water = 1 l

5. tris-Citrate system (Cardy and Beversdorf, 1984).

   (*a*) Electrode buffer (pH = 6.5)
   0.04 M tris = 4.85 g
   0.012 M Citric acid (monohydrate) = 2.50 g
   Water = 1 l

   (*b*) Gel buffer (pH = 6.5) (1:3 dilution of electrode buffer)
   0.013 M tris
   0.004 M Citric acid (monohydrate)

   POWER: 20 Watt, 230 V × 65 mA; duration = 6 hr

   GEL: 12.5% w:v starch = 64 g
   3.0% w:v sucrose = 15 g

6. tris-Citrate/histidine system (Cheliak and Pitel 1984).

   (*a*) Electrode buffer (pH = 7.0)
   0.125 M tris = 15.13 g
   (pH with 1.0 M citric acid)

   (*b*) Gel buffer (pH = 7.0)
   0.05 M Histidine-HCl = 10.48 g
   1.4 mM EDTA = 0.47 g
   (pH with 1.0 M tris)
   Dilute 1:4 gel buffer to water before use

   POWER: 270 V × 75 mA, Duration = 6 hr

7. tris-Citrate-histidine system (Tanksley 1979).

   (*a*) Electrode buffer (pH = 7.0)
   Citric acid = 9.04 g
   0.13 M tris = 16.35 g

   (*b*) Gel buffer (pH = 7.0)
   0.007 M Histidine = 1.05 g
   [adjust pH with NaOH]

8. **Histidine-citrate system (Cardy and Beversdorf 1984).**

   (*a*) Electrode buffer (pH = 6.5)
   0.065 M L-Histidine = 10.09 g
   0.007 M Citric acid (monohydrate) to adjust the pH = 1.50 g

   (*b*) Gel buffer (pH = 6.5) (1:3 dilution of electrode buffer)
   0.016 M L-Histidine
   0.002 M Citric acid (monohydrate)

   POWER: 275 V × 47.5 mA, Duration = 6 hr.

9. Histidine-citrate system (Fildes and Harris 1966).

   (*a*) Electrode buffer (pH = 7.0)
   0.005 M Histidine = 0.776 g
   [adjust pH with 7.0 NaOH]

(*b*) Gel buffer (pH = 7.0)

0.41 M Sodium citrate (trisodium)	= 105.81 g
[Adjust pH with citric acid—anhydrous]	= 78.76 g
Water	= 1 l

10. tris-Glycine system (Scandallios 1969).

(*a*) Electrode buffer (pH = 8.75)

0.02 M tris	= 3.09 g
0.19 M Glycine	= 14.40 g

(*b*) Gel buffer is the same as electrode buffer

11. tris-Versene-borate system (Shaw and Prasad 1970).

(*a*) Electrode buffer (pH = 8.0)

tris	= 60.60 g
Boric acid	= 40.00 g
Na_2EDTA	= 6.00 g

(*b*) Gel buffer (pH = 8.0)

tris	= 6.06 g
Boric acid	= 6.00 g
Na_2EDTA	= 0.60 g
Water	= 1 l

12. Histidine-citrate system (Goodman et al. 1980).

(*a*) Electrode buffer (pH = 5.0)

0.05 M L-Histidine	= 7.75 g
0.024 M Citric acid (monohydrate)	= 5.00 g
[adjust pH with citric acid]	

(*b*) Gel buffer (pH = 5.0) (1:12 dilution of electrode buffer)
0.004 M L-Histidine
0.002 M Citric acid (monohydrate)

13. tris-Citrate system (C. Sneller, personal communication).

(*a*) Electrode buffer (0.2 M tris, pH = 6.65)

1 M tris	= 200 ml
Water	= 800 ml
[pH with citric acid]	

(*b*) Gel buffer (0.0075 M tris, pH = 6.65)

1 M tris	= 9.5 ml
Water	= 992.5 ml
[pH with citric acid]	

14. Phosphate system (Shaw and Prasad 1970).

(*a*) Electrode buffer (0.2 M phosphate, pH = 5.8)

$NaH_2PO_4.H_2O$	= 27.8 g
$NaH_2HPO_4.7H_2O$	= 53.65 g

(*b*) Gel buffer (pH = 5.8)
Add 50 ml of electrode buffer to 1000 ml of water.

15. Phosphate system (Shaw and Prasad 1970).

(*a*) Electrode buffer (pH = 7.0)

K_2HPO_4 (anhydrous)	= 87.0 g
KH_2PO_4 (anhydrous)	= 68.0 g

(*b*) Gel buffer (pH = 7.0)
Add 100 ml of electrode buffer to 1000 ml of water

16. Phosphate-citrate system (Shaw and Prasad 1970).

(*a*) Electrode buffer (pH = 7.0)

0.214 M K_2HPO_4 (anhydrous)	= 29.1 g
0.027 M Citric acid (monohydrate)	= 5.7 g

(*b*) Gel buffer (pH = 7.0)

K_2HPO_4 (anhydrous)	= 1.06 g
Citric acid (monohydrate)	= 0.254 g

17. Acetate system (Shaw and Prasad 1970).

(*a*) Electrode buffer (pH = 4.6)

0.2 M Sodium acetate	= 27.2 g
Acetic acid	= 12 ml

b) Gel buffer (pH = 4.6)
Add 100 ml of electrode buffer to 1000 ml of water.

pH is one of the most frequently and readily manipulated properties of buffer solutions. Table 3-1 shows the corresponding buffer compositions commonly used in electrophoresis. The pH increments are not consecutive but show three classes of buffers: acidic, neutral, and basic. It may be desirable, especially when a new species is being surveyed for the first time, for a researcher to choose from the three classes, because enzymes differ in their pH optima. With time and further tests, the pH could be fine-tuned for a specific crop or species and for the enzymes of interest. Generally, histidine buffer systems are acidic whereas lithium-borate and tris systems tend to be basic (Table 3-1). Citric acid is used at both acidic and basic pHs. Usually, one can find a general purpose buffer system (one that is able to resolve nearly all enzymes of interest); however, a second or third system may be necessary for "problem" enzymes.

Table 3-1. pH Range and composition of some electrophoresis buffers in use, with examples of crops/species.

pH	Buffer system	Source	Crop/species
5.0	Histidine-citrate	Cardy et al. (1981)	Maize
5.7	Histidine-citrate	Cardy et al. (1981)	Maize
6.1	Morpholine-citrate	G. Acquaah (unpublished)	Soybean
6.5	Histidine-citrate	Cardy and Beversdorf (1984); G. Acquaah (unpublished)	Soybean Sugar beet, wheat
7.0	Histidine-citrate	O'Malley et al. (1979)	Pine
7.5	Borate/tris-citrate	Mitton et al. (1979)	Pine
7.8	Borate/tris-citrate	Quiros (1981)	Alfalfa, medicago
8.0	tris-glycine/ tris-HCl	Sheen (1972)	Tobacco
8.1	Lithium-borate/ tris-citrate	G. Acquaah (unpublished)	Dry beans
8.3	Borate/ tris-citrate-borate	Bringhurst et al. (1981)	Strawberry
		Wall (1968)	Phaseolus
8.5	Lithium-borate/ tris-citrate	O'Malley et al. (1979)	Pine
8.75	tris-glycine	Scandalios (1969)	Pea, maize

[a]A slash indicates a discontinuous buffer system.

3.5. Preparation of Starch Gel

Starch gels are almost exclusively run in the horizontal position. Running them in the vertical position is also possible (Takacs and Kerese 1984), but this is seldom done. The starch used must be properly hydrolyzed for a high degree of resolution (Smithies 1955). Commercially hydrolyzed starch may be purchased for electrophoresis. Alternatively, some researchers prefer to hydrolyze their own starch.

3.5.1. Gel Composition

Starch gels are described according to the percentage of starch they contain (e.g. 10, 11, or 12%, etc.). The percentage is determined on a weight-to-weight (w:w) basis of starch to buffer. The percentage frequently used is between 10 and 12.5%. The lower the percentage of starch, the lighter the slices of gel will be and, consequently, the more delicate the handling required to prevent tearing of the gel. Ramshaw et al. (1980) reported that softer, less concentrated, gels enhanced the resolving power of electrophoresis; however, slices of double thickness may be needed to obtain better staining and resolution of protein bands. Gels sometimes may be supplemented with chemical compounds (such as sucrose, cofactors) for enhanced resolution (Brewer and Sing 1970). The white starch powder has to be cooked to produce a clear gel of uniform consistency. Various chemical companies offer starch (potato) labeled "hydrolized for electrophoresis." Other varieties of starch are Electrostarch™ and Connaught™ starch (light weight), both named after their respective companies. Researchers' preferences for the variety of starch vary. Some researchers use a combination of two varieties of starch for preparing gels: for example, 51.8 g Connaught™ starch and 14.9 g Electrostarch™ in 500 ml buffer (Cardy and Beversdorf 1984). Different batches of starch may produce varying rates of migration of protein and resolution; some batches may contain substances that inhibit a certain enzyme activity (Shaw and Prasad 1970). Therefore, use of a single batch of starch for a particular study is recommended.

3.5.2. Methods of Starch Gel Preparation

Starch gels can be prepared by two basic methods: (a) Bunsen burner, and (*b*) microwave oven. The amount of gel to be prepared depends upon the number of slices required and the size of the gel mold. A liquid-level may be used to set up the mold in a level position before pouring the starch gel.

3.5.2.1. General Materials The materials required for preparing the starch gel are as follows:

1. measuring cylinders (different sizes)
2. funnel
3. gloves (heat resistant)
4. volumetric flask (1 l) or Erlenmeyer flask
5. filter flask (1 l)
6. beakers (100 ml or larger)
7. starch
8. gel buffer
9. tap vacuum aspiration system
10. Bunsen burner or microwave oven (large enough to hold flasks).

3.5.2.2. Bunsen Burner Method Mix the right amounts of gel buffer and starch in a filter flask. Swirl continuously over a Bunsen burner until the white starch suspension turns clear and gelatinous (Figure 3-7). The final product should be absolutely free of lumps to offer uniform gel consistency. Do not overcook the starch because this causes loss of toughness, making slices very delicate and fragile. Knowledge of when a starch gel is considered adequately cooked is attained largely by experience. Generally, when small air bubbles start to form in the gel, stop heating. Next, connect the stoppered filter flask to a tap aspirator to degas (Figure 3-8). This step must be performed quickly to avoid cooling and setting of the viscous material while in the flask. The degassing process is completed when large air bubbles begin to appear, an event that occurs in about 30 sec for a 500 ml gel. Further degassing beyond the large air bubble stage is undesirable, since the gel will actually be boiling under reduced pressure due to the partial vacuum created by the evacuation process. Swirling the starch in the initial stages of degassing may be necessary to dislodge the tiny air bubbles trapped on the walls of the flask. This initial vigorous swirling also prevents the starch gel

Figure 3-7. Preparing starch gel over a Bunsen burner flame.

from being sucked up into the aspirator tube. Pour the degassed gel immediately into a prepared gel tray or mold, taking care not to cause air bubbles to form in the process (keep the flask close to the mold and pour into the center). If air bubbles form, they must be immediately sucked out with a Pasteur pipette.

Let the gel cool completely (about 1 hr) and then wrap the gel and mold air-tight in a plastic wrap to prevent dessication and shrinking. This step is especially critical when a gel will not be used soon after being poured. A gel deteriorates progressively with age and should be used within 24 hr of preparation. (I have used three-day-old starch gels without a noticeable loss in quality of electrophoretic results, but I do not recommend this practice.) Since a gel unavoidably shrinks a little at the top upon cooling to form a slight depression, preparing just a little more (e.g. 20 ml extra) gel than needed may be desirable so that the gel heaps above the top of the mold (C. Sneller, personal communication). The very top slice of a gel is usually discarded because the excessive exposure to the environment and handling render it rough and poor in enzyme activity and consequently poor in resolution of isozymes. Some researchers prefer to place a thick slab of acrylic on the gel after pouring to obtain a flat-topped gel upon cooling. This requires care to avoid trapping large air bubbles under the slab.

Figure 3-8. Degassing starch gel by a tap vacuum aspiration system.

Starch may also be cooked on a hot plate. Swirl the gel intermittently for uniform consistency. A magnetic bar may be placed in the starch suspension to stir it while cooking.

3.5.2.3. Microwave Oven Method Pour about ⅔ of the gel buffer into a volumetric flask and bring it to a boil in a microwave oven. Mix the remainder of the buffer with the starch in a filter flask and swirl it to keep the starch in suspension. Pour the boiling buffer into the starch suspension while swirling. The long narrow neck of the volumetric flask helps to deliver the hot buffer into the starch suspension quickly and without spilling (Figure 3-9). Still, perform the above operation over a sink because occasionally the hot buffer may gush out of the filter flask. Cook the gelatinous product for a few more minutes until tiny air bubbles start to appear, then degas it as in the Bunsen burner method. The microwave oven method may also be adapted to use of a Bunsen burner or hot plate.

The duration of the various operations will vary with the amount of starch cooked and the heat required to cook it. When a 600 Watt microwave oven is used at high power, the boiling of the ⅔ volume (350 ml for a 500 ml gel) of gel buffer takes about 4 to 5 min and further cooking of the gel takes about 1.5 min.

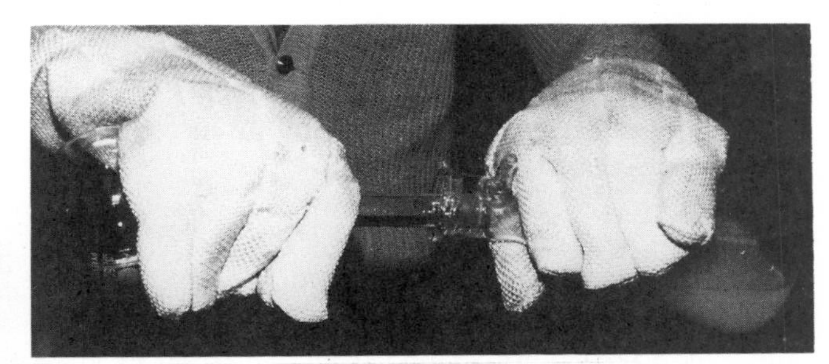

Figure 3-9. Delivering hot buffer solution into flask containing starch suspension during gel preparation.

REMINDER

1. Wear protective gear: insulated gloves, laboratory coat, safety goggles.
2. Beware while introducing recently boiled buffer into the starch suspension of the pos-

sibility of the hot buffer gushing out of the filter flask. Point the flask away from you during the operation.

3. Do not overcook the starch.
4. Do not overdegas the gel.

3.6. Sample Collection and Preparation

The type of material to sample for electrophoresis is determined by several factors, including the following:

1. Enzymes to be assayed: Some enzymes or proteins are tissue-specific (e.g. phaseolin, a seed storage protein, is found only in seed) or are better separated when certain tissues are used. In common bean, diaphorase is detected much better in root tissue than in seed or leaf. The tissue-specificity of some enzymes has been reported in some crops, for example peroxidase in *Pisum sativum* (Manicol 1966), and esterase, peroxidase, and catalase in *Zea mays* (Scandalios 1964). Shikimate dehydrogenase is not expressed in *Beta vulgaris* pollen grains (Aicher 1988).
2. Ease of sampling and homogenizing: Why spend extra time and money to raise seedlings or plants when seeds can be used instead? Furthermore, seeds can be sampled nondestructively and then planted for later data collection. It must be borne in mind, however, that certain plant tissues are softer and therefore easier to homogenize than others.
3. Developmental stage of tissue source: Some enzymes are present only at a certain developmental stage(s) in the life cycle of an organism (see 2.15) (Thurman et al. 1965).
4. Growth environment: The effects of light, temperature, and seasonal variation on enzymes are discussed above (2.15).

3.6.1. Collecting Samples

Standardize the sample collection process, and for that matter all steps in electrophoresis, as much as possible. Standardization is critical because electrophoretic analysis is largely comparative. Because enzymes are influenced by a host of factors, discrepancies in methodology can produce a false basis for declaring differences among subjects. Use equal sizes or amounts of tissue and equal amounts of extraction buffer so that the concentration of the various extracts being investigated will be uniform. When taking samples from leaves, a cork borer may be used to cut out equal sizes and numbers of leaf disks. The sample should be representative of the source. Since tissue age is a factor in the activity of some enzymes, the leaves picked should be of equal age. Some enzymes are more heat-labile than others but it is a good practice to keep all samples cold throughout the sampling process and thereafter, especially if the samples are collected well before they are to be prepared. Samples should be kept on ice if they are collected from the field or outside the laboratory.

Fresh samples are therefore preferred for electrophoresis. Nonetheless, when large numbers of samples are to be analyzed, preparing the samples ahead of time may be more convenient and practical. The second option requires some kind of storage. Conventional freezer storage (< -20°C) may suffice for storage duration of a few days and sometimes a few weeks, but for long-term storage, a super cold freezer (-80°C) is a better alternative. Generally, repetitive freezing and thawing is not recommended for biological materials because this predisposes them to rapid deterioration. Remember that once a sample is taken from a parent source, the quality of the tissue can only deteriorate. Cells begin to die and enzymes become progressively less active. Therefore, storing a whole tissue would be better than storing a homogenized (processed) one. If fresh materials are not to be used immediately, they must be frozen rapidly, as is generally recommended for biological materials. Stored samples may undergo biochemical changes that can simulate polymorphism as a result of unstable compounds changing their chemical state (Walter et al. 1965; Lai 1966).

Seed tissue samples may be obtained from dry seed (by scraping) or after seeds have imbibed water for a few hours under dark and cool conditions (proteins may deteriorate if seeds are soaked for a long period). Containers with separate small wells or petri dishes lined with absorbent paper may be used for soaking seeds in water. Cotyledons from seeds that have imbibed water are easier to grind than dry seed. When nondestructive sampling is required, a piece of cotyledon may be cut (or scraped in the case of dry seed) from the raphe end of the seed so that the remainder can be planted. Seedlings may be raised in vermiculite so that the roots can be readily sampled by uprooting.

REMINDER

1. "Garbage in, garbage out!" Use the best samples.
2. Keep samples cold throughout the collection period.
3. Record the "sample specifics" (age, amount, source, etc.). This information is helpful in data intepretation.

3.6.2. Sample Preparation

Manual and motorized are the two basic procedures used to homogenize samples. The procedures require the following materials:

Manual (Figure 3-10)	Motorized (Figure 3-11)
porcelain spot plate	motorized unit
pestle	microcentrifuge tubes
tray packed with ice	tube rack on ice.

In addition, the following items are common to both methods: extraction buffer and Pasteur pipette or pipette dispenser, specimens, wicks, and forceps.

One way of homogenizing plant samples is by using a porcelain mortar and pestle. The mortar can be in a form of a spot plate (Figure 3-10). When using a spot plate, grind slowly to avoid contamination from splashing of extracts from one spot to another. Add the proper amount of the extraction buffer to the samples before grinding. The ground sample should have the consistency of a slurry. The grinding stage is critical because too much buffer will produce a very dilute extract, which in turn may result in faint or no bands upon staining. Samples taken from leaves with a cork borer of diameter 1 cm may be homogenized with five drops of extraction buffer per three disks. A Pasteur pipette may be used to deliver the buffer. It is better to add a little less buffer to a sample when homogenizing it so that more can be added later as needed. For seeds soaked in water, a 100 mg of tissue may require about 25 μl of buffer. The researcher should determine the best ratio of tissue:buffer for a particular situation by trial and error. The grinding should be done while the plate sits on ice to keep the materials cold and thereby preserve the activity of very heat-sensitive enzymes.

Figure 3-10. Apparatus for manual homogenization of samples for electrophoresis: (*a*) extraction buffer and dispenser;(*b*) beaker of water; (*c*) paper towel; (*d*) wicks;(*e*) forceps; (*f*) pestle and spot plate on ice.

With motorized grinding (Figure 3-11), the sample is placed in a microcentrifuge (Eppendorf) tube (1.5 ml) along with the appropriate amount and type of grinding buffer and ground with a motor-driven

Figure 3-11. Apparatus for motorized homogenization of samples for electrophoresis: (*a*) motor; (*b*) rotating pestle; (*c*) container in which pestle is washed; (*d*) water bottle; (*e*) paper towel; (*f*) microcentrifuge pipette rack on ice.

pestle. Pushing and rotating the tube aids the grinding process. The homogenized sample may be centrifuged to obtain a supernatant. The centrifuge may be placed in a refrigerator to keep the samples cold, since the equipment generates some heat during operation. Do not miss an opportunity to keep things cold.

After homogenizing the samples, paper wicks (cut from suitable paper, such as Whatman paper, No. 2 or 3, chromatographic paper) are used to soak up the extract. The paper is cut into narrow strips, of suitable dimensions (e.g. 2 mm × 12 mm, 4 mm × 12 mm, etc.), and should be slightly longer than the thickness of the gel in which they will be inserted for electrophoresis so they can readily be removed (dewicked) if so desired. Wick dimensions may be changed to suit specific purposes. Feel free to experiment with different sizes of wicks. If necessary, wicks soaked in extracts may be stored for a very short period just as described for fresh samples. Soak duplicates of wicks in each specimen extract so that one set of wicks can be stored for use in repeat runs in the immediate future, if necessary. When the motorized method is used for grinding samples, the microcentrifuge tubes may be saved for the repeat tests.

REMINDER

1. Wear protective gear (at least gloves) when grinding with buffers that include harzardous additives such as 2-mercaptoethanol.
2. Grind in a fume hood or wear a dust mask to avoid inhaling volatile and toxic buffer components.
3. Be careful to use correct buffer:sample amounts for uniform concentration of extracts.

3.7. Extraction (Homogenization) Buffers

Tissues are homogenized in a buffer to liberate soluble enzymes for electrophoresis. The quality of extract, among other factors, is critical to the quality of results of electrophoresis. In some tissues, the presence of quinones and phenols may adversely affect the quality of an extract (Wilson and Hancock 1978). These organic compounds have oxidizing properties and may deactivate certain enzymes (Loomis and Battaile 1965). Extraction buffers therefore frequently include anti-oxidants and other reagents that help to preserve enzyme activity during extraction. Wilson and Hancock (1978) compared the effects of some of these additives, namely, polyvinylpyrolidone (PVP), anion exchange resin, cystein hydrochloride, and ascorbic acid, and found that enzymes respond differently to buffer composition, as evidenced by the differences in the quality of resolution of enzymes on a zymogram. Anderson (1968) added bovine albumin (40 mg/100 ml) to the extraction buffer to bind up phenolics and free fatty acids for better resolution. The addition of inhibitors such as ethylenediamine tetraacetate (EDTA) to buffer solutions suppresses or eliminates the expression of undesirable volunteer bands that complicate scoring of gels (Hames 1981). Harris and Hopkinson (1976) recommended the addition of the appropriate cofactor to the extraction buffer and/or gel buffer to help stabilize an enzyme for more consistent results.

Some extraction buffers are suited to a wide variety of enzymes. What works for one species, however, may not work as well for another. Examples of extraction buffers are presented below. The basic ingredients in the original recipes generally remain unchanged, but some of the pH requirements and additives have been modified by various researchers. When working with a new species, one may have to try several buffers before adopting one or two as standard. Some buffers are easier to prepare and are less expensive than others. Other buffers may include ingredients that are dangerous to health and hence require special care when handled.

3.7.1. Selected Recipes for Extraction Buffers for Plant Materials

1. Phosphate buffer (Mitton et al. 1979), pH = 7.5.

BUFFER: 0.10 M (KH_2PO_4)

0.029 M Sodium tetraborate	= 0.28 g
0.20 M L-Ascorbic acid (Na salt)	= 1.0 g
0.016 M Diethyldithiocarbamic acid (Na salt)	= 0.07 g
PVP–40 T	= 1.0 g
2-Mercaptoethanol (add just before use)	= 1%
Water	= 25 ml

SPECIES/CROPS: e.g. fern

2. tris-Maleate buffer (Soltis et al. 1983), pH = 7.5.

0.020 M Sodium tetraborate	= 1.91 g
0.02 M Sodium metabisulfite	= 0.10 g
0.25 M L-Ascorbic acid (Na salt)	= 1.24 g
0.026 M Diethyldithiocarbamic acid (Na salt)	= 0.11 g
0.10 M Maleic acid	= 0.29 g
0.10 M tris	= 0.30 g
PVP–40 T	= 1.0 g
Water	= 25 ml
(pH with 1.0 M HCl)	
2-Mercaptoethanol (add just before use)	= 0.03 ml

SPECIES/CROPS: e.g. fern, dry beans

3. Sucrose buffer (Cardy et al. 1981), pH = 7.38.

16.7% Sucrose (w:w)	= 4.09 g
8.3% Na ascorbate (w:w)	= 2.03 g
Water	= 25 ml

SPECIES/CROPS: e.g. maize, soybean, wheat, barley

4. tris-HCl buffer (Gottlieb 1981), pH = 7.5.

0.010 M KCl	= 0.019 g
0.001 M EDTA (tetrasodium salt)	= 0.01 g
0.1% 2-Mercaptoethanol	= 0.025 ml
0.01 M $MgCl_2.6H_2O$	= 0.05 g
or PVP–40 T	= 1 g
0.10 M tris-HCl	= 25 ml

SPECIES/CROPS: e.g. fern

5. tris Buffer (Bringhurst et al. 1981), pH = 7.5.

0.01 M tris	= 30 mg
0.10% Cystine-HCl (w:v)	= 0.4 g
0.10% Ascorbic acid (w:v)	= 0.4 g
0.10 gm/ml PVPP	= 2.5 g
Water	= 25 ml

SPECIES/CROPS: e.g. strawberry

6. tris Buffer (Hauptli and Jain 1978), pH = 7.0.

0.1 M tris	= 300 mg
0.14 M 2-Mercaptoethanol	= 0.3 ml
Water	= 25 ml

SPECIES/CROP: e.g. amaranths

7. tris-Maleate buffer (Weeden and Emmo, n.d.), pH = 8.0.

tris	= 1.21 g
Water	= 80 ml
(pH with maleic anhydride)	
Glycerol	= 10 ml
(hydrate overnight or stir into solution)	
PVP–40 T	= 10 g
(hydrate overnight or stir into solution)	
Triton X-100 (add just before use)	= 200 μl
2-Mercaptoethanol	= 10 μl

SPECIES/CROPS: e.g. common bean, sugar beet, barley

8. Phosphate buffer (original source not found), pH = 7.5.

EDTA	= 45 mg
DTT (Dithiothreitol)	= 38 mg
0.1% 2-Mercapthoethanol	= 25 μl
0.5% Tween–80	= 125 μl
(adjust pH to 8.0 with KOH)	

SPECIES/CROP: e.g. blueberry

9. Water (Schwartz 1960).
Used for esterase

SPECIES/CROP: e.g. maize

10. 0.9% w:v NaCl (Scandalios 1967).
Used for alcohol dehydrogenase

SPECIES/CROPS: e.g. maize

11. Gluthione buffer (Arus and Orton 1983), pH = 8.5.

0.10 M tris = 300 mg
1.0% Reduce gluthione (w:v) (pH with 3 N NaOH)
Water = 25 ml

SPECIES/CROP: e.g. cabbage

Grinding or extraction buffers may be prepared in bulk and stored in a refrigerator until needed. Additives such as triton X-100 and 2-mercaptoethanol should be added just before use. When using 2-mercaptoethanol, grind the samples under a hood.

3.7.2. Recipes for Extraction of Animal Material

1. tris-Citrate buffer (Correa-Victoria 1987), pH = 8.7.

0.1 M tris-Citrate	= 1000 ml
Sucrose	= 170 g
Ascorbic acid	= 1 g
Cysteine HCl	= 1 g

ANIMAL TISSUE: fungi

2. tris Buffer (Feder et al. 1989), pH = 7.0.

0.1 M tris
EDTA
0.13 mM $NADP^+$
0.15 mM NAD^+
0.3 M 2-mercarptoethanol

ANIMAL TISSUE: insects (e.g. butterfly, maggot fly)

Solution 2 can be used for blood samples. Blood is collected in heparinized tubes. A drop of heparin (1 g/100 ml water) is added to the blood sample and mixed and centrifuged to separate the plasma from the red blood cells. The red blood cells are lysed with water (1:1). Samples must be stored at supercold temperature (−80°C).

3.8. Loading a Gel for Electrophoresis

One may have the best quality of enzyme extracts and still ruin the outcome of electrophoresis by not properly loading the samples into the starch gel. Loading is the stage at which wicks that have been soaked in sample extracts are inserted into the gel for electrophoresis. Each of the two models of gel mold has a convenient way in which it may be loaded with sample wicks.

3.8.1. Materials for Loading a Gel

1. A pair of forceps
2. A straight edge
3. Scapel or knife

3.8.2. Methods of Loading a Gel

Prior to loading, the gel may be placed in a refrigerator to cool for about 20–30 min. First, a straight, smooth cut is made in the gel to the bottom of the mold. A wavy slit may cause uneven movement through the gel and produce inaccurate relative mobilities of the protein bands. The position of the cut depends upon the expected direction of migration of the enzymes being assayed. If only anodal (positive pole) migration is expected, a cut positioned at about 3 cm (ruler's width) from one end of the gel is adequate, otherwise, the cut may be made a little further toward the middle of the gel. Sometimes unsatisfactory migration of proteins may be caused by loading of samples too close to the sponge.

To load the gel, pick up one tip of the wick with a pair of forceps and soak it in the appropriate sample extract. Blot away any excess extract by placing the wick on an absorbent material (e.g. tissue paper). To avoid soaking up debris from the crude extract into the wick, some researchers prefer to soak up the extract through e.g. a 35 μm nylon mesh (Krebs and Hancock 1988). Next, insert the wick into the slit by opening it with the aid of two fingers, pushing the gel on either side of the slit. Sometimes the slit may be opened with just one finger pulling on one side of the gel (Figure 3-12). The above procedure applies to gels cast with a Type I model mold with fixed walls and the Type II model. The wick should be pushed down so it touches the bottom of the mold and must be in an upright position (do not slant the wicks!).

Figure 3-12. Loading a Type I (rigid walls) or Type II gel mold for electrophoresis.

When loading a gel, try to space the wicks evenly in the gel, making certain that adjacent wicks do not touch. A minimum space of 1 mm between adjacent wicks is recommended. Should one choose to use a thin wick, loading the gel while holding onto one end of the wick may be difficult. In this case, the wick may be held in the forceps, and the forceps inserted into the slit and pushed down until it touches the bottom of the form. By slightly releasing one's grip on the forceps, the wick is freed into the slit.

Initially, even spacing of the wicks may be difficult to achieve. For those using a Type I mold with loose walls, it is more convenient to free the wall at the cathodal end of the gel after making the cut so that the cathodal strip of gel can be moved back to give enough room for loading. The loading is then accomplished by pasting the wicks on one cut surface (Figure 3-13). The cathodal piece (smaller) is later pushed back into position and the wall taped down before electrophoresis. Placing a marked strip of paper or ruler near the slit may be helpful (Figure 3-13). Avoid loading the wicks too close to the edge of the mold, since the migration front may sometimes curve up at the edges. The presence of wicks in the slit traps pockets of air, which impede the flow of electrical current through the gel. A strip of acrylic or drinking straw (or any suitable material) may be firmly inserted between the gel and the wall of the mold at the cathodal end of the mold to help close up the slit. This must be done carefully, especially when using a Type II mold, to prevent causing a break in the gel in the vertical parts of the mold. Sometimes the slit can be closed by just pushing the two sections of the gel together.

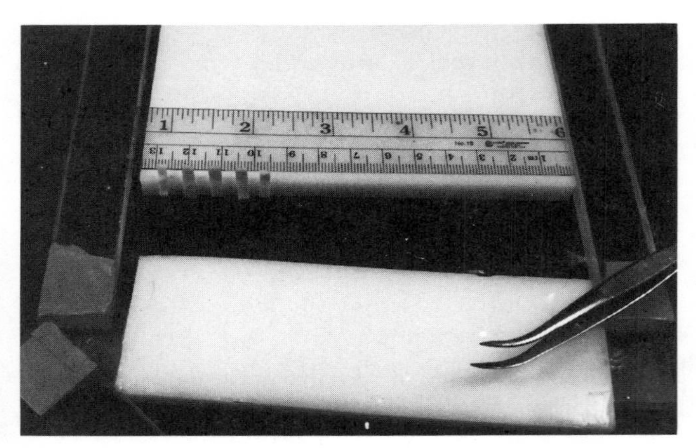

Figure 3-13. Loading a Type I gel mold with loose walls for electrophoresis.

As much as possible, a gel should be loaded in a balanced way so as to avoid distorting the migration of proteins (Brewer and Sing 1970). Balanced loading means that the wicks should not only be evenly spaced in the gel but should also cover the width of the gel. Empty spaces may be filled with dummy, duplicate, or dyed wicks. When working with replicated samples, one should arrange the specimens strategically to help in scoring (or "reading") the stained gels. Strategic arrangement of samples is necessary because scoring a gel is primarily a comparative operation.

Some kind of "design" is necessary in gel loading. Figure 3-14 shows some of the designs possible to employ in loading a gel whereby each sample has four replicates. *Complete-blocking* on the basis of some a priori information on the materials being investigated (e.g. same treatment, same cultivar, same plant, animal or organ, etc.) may be desirable in some studies. A variation of the complete-blocking design is the *split-blocking* design, which places replicates of the same sample on either half of the gel. The *set-blocking* pattern ensures that each entry is represented in all sec-

```
a). "Complete blocking"
==================================================
M A A A A B B B B C C C C D D D D M E E E E F F F F
==================================================

b). "Split blocking"
==================================================
M A A B B C C D D E E F F G G M A A B B C C D D E E
==================================================

c). "Set blocking"
==================================================
M A B C D E F G A B C D E F G A B C D E F G M A B C
==================================================
```

Figure 3-14. Some designs for loading wicks into gels for electrophoresis. A–G represent different entries; M is the wick with the marker dye.

tions of the gel. The advantage of set-blocking is that an entire treatment is not jeopardized if one section of the gel is damaged (e.g. tearing, warping, irregular front), as would be the case in the complete-blocking pattern.

When investigating pure lines and heterozygotes, it may be desirable to sandwich heterozygotes between homozygotes for easy comparison. Choices should be judicious and strategic. For example, in work to confirm an observation of polymorphism, retesting an entire set of samples may not be necessary. The samples with the putative variants, along with some standard (control) genotypes, should be included. With the a priori knowledge of putative polymorphism, samples then should be arranged so as to confirm the polymorphism unambiguously.

As in any other experiment, control or check genotypes should be included in electrophoretic studies where appropriate. The check genotypes are those whose isozyme patterns have been conclusively established from previous studies. To indicate the migration of charged particles through the gel and to determine the direction and distance of migration of the gel front for the purpose of estimating the relative mobilities of isozymes, insert a dyed wick (usually in the first or middle lane) in the gel. Dyes have different directions of migration in an electric field. Bromophenol blue may be used to detect anodal migration whereas methylene blue may be used for cathodal migration. The loaded gel may be covered with a piece of plastic wrap to avoid direct contact between the gel and the cooling bag, which may be placed on it for additional cooling during electrophoresis. Without a plastic cover, the slit may open during electrophoresis due to drying of the cut edges of the gel.

REMINDER

1. Do not freeze the gel when cooling it prior to loading.
2. Avoid contact between wicks; space properly.
3. Squeeze out air and close the slit (origin) firmly.
4. Blot away excess extract from wicks and from the slit.
5. When using frozen wicks already soaked in sample extract, allow the wicks to thaw; blot away the excess moisture before loading.
6. Keep wicks erect in the gel and pushed to the bottom of the mold.

3.9. Running a Gel

Running a gel is the only stage in electrophoresis that directly requires electricity. Various steps of the procedure hold the potential danger of an electrical shock. Running a gel entails placing a loaded gel in an electric field to cause molecules with different net charges to migrate in directions appropriate and at rates proportional to each one's net charge or molecular weight, respectively. Running a gel is the most time-consuming step in SGE, requiring from 3 to 18 hr. The duration of the run depends on the buffer system, the thickness of gel, and the rate chosen by the researcher.

An electrophoresis buffer system is developed on either one of the two types of gel mold and hence has unique characteristics with respect to electrical parameters. When a Type I-based (developed on a Type I mold) buffer system is run on a Type II-mold system, power settings (volt × amps) different from those recommended by the developer of the buffer system might be required (e.g. instead of 50 mA × 200 V, the meters on the power pack might register something like 120 mA × 75 V when one tries to increase the voltage to the recommended level).

3.9.1. Materials for Running a Gel

1. Power supply (DC)
2. Electrophoresis apparatus (buffer tanks, gel mold loaded with samples)
3. Electrode buffer
4. Water bag (optional)
5. Ice pack (optional)
6. Refrigerated environment (or suitable cold environment)

3.9.2. Setting up Apparatus for Starch Gel Electrophoresis

The steps involved in setting up apparatus for electrophoresis depend upon the type of gel mold model used. The most critical difference between the two models (i.e. with or without a wick) is most evident at this stage.

3.9.2.1. Method A: For Type I Gel Mold Model The buffer tanks should be filled to about ⅔ of their volume with the appropriate electrode buffer. Place the loaded gel across the tray and set up the sponges as shown in Figure 3-15. The sponges should have straight edges that are parallel to the origin of electrophoresis. Uneven sponge edges may cause an uneven gel front and consequently produce false mobility differences among isozymes.

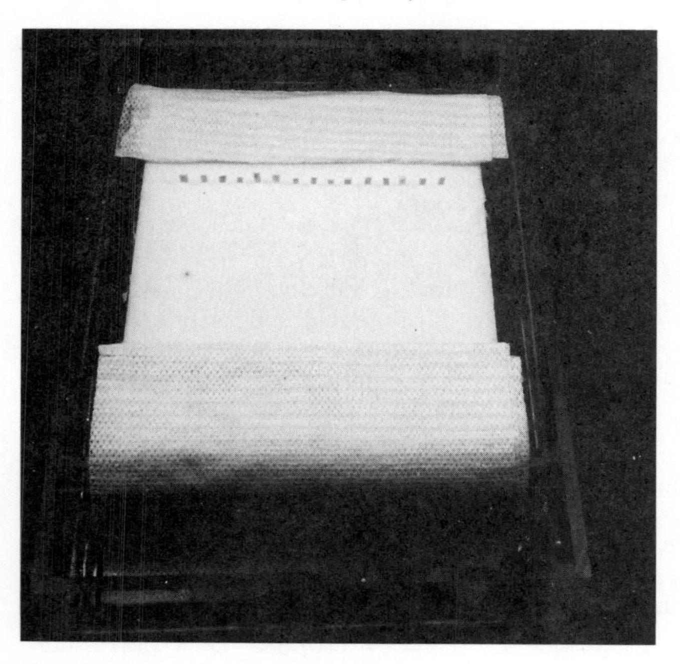

Figure 3-15. A Type I electrophoresis unit with gel loaded with sample wicks and installed for electrophoresis.

3.9.2.2. Method B: For Type II Gel Mold Model Fill up the buffer tanks as in method A above. The masking tape must be removed before setting the gel mold in the buffer tanks (Figure 3-16). Be certain that the exposed gel in the "legs" of the gel mold is completely covered with electrode buffer. When the gel mold is in place, connect the power cords to the appropriate terminals.

3.10. Further Comments on Running a Gel

Gels are usually electrophoresed from the cathode or negative (black) terminal to the anode or positive (red) terminal. When so arranged, the wick end of the gel is at the cathodal end of the electrophoresis apparatus.

Figure 3-16. A Type II electrophorsis unit showing a water bag and ice pack for supplemental cooling of the gel during electrophoresis.

Electrophoresis is conducted in a refrigerated or cold environment, especially when native gels are being run, to preserve the physiological integrity of proteins. As in other aspects of electrophoresis, the researcher can be very creative in devising a cold environment. A conventional refrigerator offers a very simple source of cold environment for electrophoresis (Figure 3-17). A homemade cooling system used in Prof. J. Hancock's laboratory at the Michigan State University is shown in Figure 3-18. The homemade cooler is constructed from a picnic box and uses running tap water and crushed ice. Keeping the gel cold throughout electrophoresis may require, in addition to a refrigerated environment, placing a water bag and an ice pack on the gel.

Fast runs generate heat quite rapidly. Check for heating by touching the gel. (Turn off the power supply before touching!) The wicks may be left in the gel throughout the entire run or removed after about 20 to 30 min. Removal of the wicks has been reported to improve band definition (Soltis et al. 1983). Some buffer systems require different levels of power for the beginning (loading run) and the remainder of electrophoresis.

Wicks, especially when cut from thick paper, create air pockets in the slit into which they are loaded. If the wicks are removed during electrophoresis, one must be sure to close the slit firmly to remove the trapped air. This is done by applying pressure with the fingers to the gel on both sides of the slit and pushing together. Trapped air interferes with the flow of electricity through the gel. Acrylic plastic strips or drinking straws may be inserted, as described previously. Improper closing of the slit may cause a wavy gel front and pose a problem in the scoring of the zymogram. The slit may be wiped clean of any remaining extract after removing the wicks by running a cotton swab (or any suitable material) through it a few times.

If the source of power supply to be used is not in good working condition or if the apparatus is set up improperly for electrophoresis, unstable electrical parameters may occur during electrophoresis.

Figure 3-17. Using a conventional refrigerator to provide cooling during electrophoresis.

Figure 3-18. A homemade cooling system for electrophoresis. (Dr. J. Hancock, Jr., Department of Horticulture, Michigan State University.)

Changes in current and voltage may also be due to ionic fluctuations of the gel and electrolyte (due partly to electrolysis at the electrodes). Check the power setting periodically to avoid the possible heating of the gel, which would endanger the more thermolabile

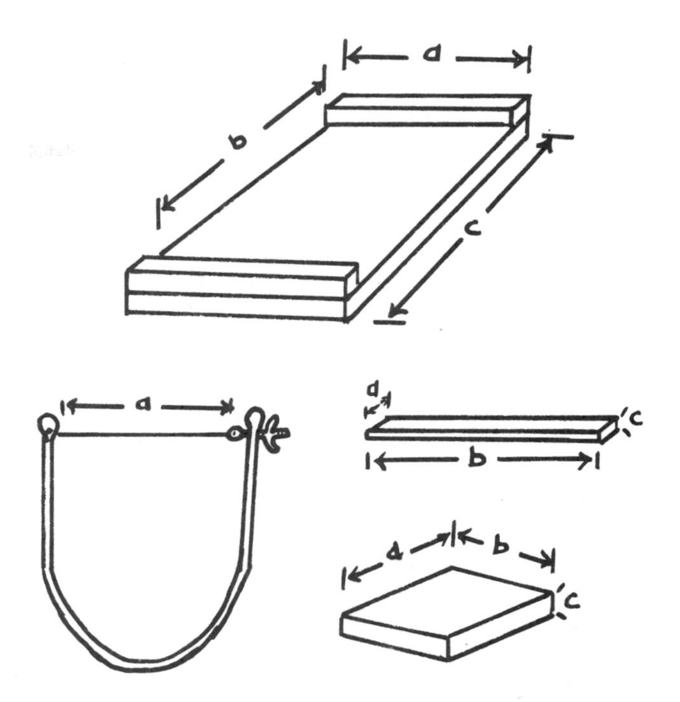

Figure 3-19. Apparatus for slicing a gel after electrophoresis: (*a*) slicing bed ($a = 19$ cm, $b = 25$ cm, $c = 28$ cm); (*b*) spacer ($a = 1.5$ cm, $b = 25$ cm, $c = 1$ cm); (*c*) slicer ($a = 25$ cm); (*d*) slicing slab or weight ($a = 17$ cm, $b = 13$ cm, $c = 1$ cm).

enzymes or cause the gel to run too slowly, thereby wasting time. A reliable power pack is required for overnight, unattended electrophoresis.

Generally, electrophoresis is completed when the tracking dye has migrated about 10 cm from the origin. The dye front is always ahead of the actual locations of the sample molecules in the gel. Insufficient duration of electrophoresis may lead to poor separation of isozymes and problematic scoring of the gel. Fresh electrode buffer solution may be used for each electrophoresis, but most buffers may be used two to three times before changing. Since electrophoresis is conducted under high-voltage conditions, post a caution sign reading "danger high voltage" on the power packs or refrigerator to warn visitors of the potential hazard.

REMINDER

1. Provide a cold environment for electrophoresis but do not freeze the gels.
2. Connect the terminals correctly.
3. When using supplemental cooling aids (ice block and water bag), make sure the surface of the gel is evenly covered for uniform cooling.
4. Keep sponges saturated with buffer when using a Type I mold. For a Type II mold, be certain the exposed gel in the legs is submerged in the electrode buffer.

5. Check the gel periodically to ensure the proper voltage and current. If two gels are connected to one power pack that delivers constant current, set the current to two times the amount for one gel if the two gels are controlled by the same knob on the power pack. The voltage will be the same as for one gel.
6. Do not handle the gel until the power supply has been turned off if you want to live to stain the gel yourself!
7. Recheck to make sure that all power cords are properly connected and that the desired current is being supplied. Even very experienced users of electrophoresis make mistakes every now and then! Nothing, perhaps, is more annoying than to come back after six or more hours of electrophoresis to find that an irreverrsible mistake has rendered the whole day's work useless!

3.11. Preparing Gel Slices for Staining

After electrophoresis has been completed, the power must be turned off before beginning to unload the gel to prepare for staining. When using a Type I mold with loose wall pieces, the gel is exposed by simply removing the tape (or paper clamps) to free the walls of the gel mold. There is no need to lift the gel from the mold because the bed of the mold can serve as a slicing bed. In the Type II system and the Type I system with fixed walls, the gel is cut out of its mold. The portion of the gel above the marker dye front is cut away because to include the top, "blank" part of the gel wastes staining solution. Slide the scapel or knife around the other sides of the gel to free them from the walls of the mold. Alternatively, one may also cut out strips of "blank" gel from the sides if the wicks were not loaded too closely to the edges for electrophoresis. The top left corner of the gel is usually sliced off to indicate the starting point of loading of samples (i.e. if loading was started from the left to right).

Remove the gel from the mold by turning the mold upside down into the palm of the hand and gently prying the gel from the bottom of the mold, beginning at one of the edges. When the gel is freed into the palm, place it on a slicing bed (Figure 3-19). It may be easier to place the slicing bed on the gel while the latter is still being held in the palm, then flip over and position the gel properly on the bed. Trapped air may be removed by pressing down on the gel with gentle sweeping motions toward the edges. Sometimes, the gel may stick to the bottom of the mold and prove difficult to remove without breaking it. When that happens, draw a thin string between the bottom of the gel and the bed as though slicing it. Alternatively, slide a smooth, wet, flexible spatula under the gel. To minimize sticking, use very clean gel molds.

A sheet of very thick acrylic or glass may be placed on the gel to keep it steady during slicing. Stack the slicing strips or spacers of thin acrylic on both sides of the gel while drawing a string [e.g. nylon, E (6th) guitar string], either hand-held or strung between a U-clamp (Figure 3-19), smoothly through the gel, all the while pressing down on either side to ensure that the strips are flat for even thickness of the slices (Figure 3-20). The strips may be wetted with water to make them stick to each other as they are piled up for slicing. That equal numbers of spacers are placed on either side of the gel before slicing is critical. It is also important to make sure the string used for slicing is very clean and smooth before each slicing operation. Wipe the string clean after a couple of slices are cut and certainly after it has been left unused for a while. A rough string will produce slices with an uneven surface, which may affect the intensity of bands. Stain the sliced gels following the steps in Chapter 2.

Figure 3-20. Slicing a gel.

3.12. Gel Fixation

As mentioned above, some bands fade away quickly, yet prolonged staining can produce overstained bands and consequently unscorable zymograms. Even after the staining solution has been drained out, some bands continue to increase in intensity while others fade away. To arrest the staining process completely and preserve the bands in their peak form, the gel must be fixed in a special solution. A general purpose fixative is 5:5:1 solution of methanol:water:glacial acetic acid (Cardy and Beversdorf 1984). Gels stained with solutions containing MTT are fixed much better in glycerol:acetic acid:water:ethanol at 1:2:4:5 (Stuber et al. 1988). Other fixatives include 50% ethanol or 1% glacial acetic acid. If a fixed gel fades away quickly, a 50% glycerol fixative may be used. After fixing, some of the background color disappears, making the bands more clearly visible.

3.13. Photographing Gels

Gels may also be photographed for various purposes. A light box (e.g. physicians x-ray reading light box) helps in scoring gels and in photographing them. Photographic films must be chosen to meet the lighting conditions. Slide films are preferred because they can be projected and enlarged for detailed studies of patterns. For the purpose of publication, the researcher may invest in high-quality gels by following the recommendations of Thorpe et al. (1989), who advise adding the appropriate coenzyme to the gel buffer (about 10 mg/100 ml) before preparing the gel, and similarly to the tank buffer before electrophoresis. The purpose of this action is to ensure that most of the coenzyme-binding sites of the enzyme are occupied so that its activity will be restricted to a discrete zone. Other quality-enhancing techniques for zymograms include increasing the gel concentration and extending the duration of electrophoresis for better separation.

3.14. Photocopying Gels

A hard copy of a zymogram can be obtained by simply photocopying the gel. The gel should be placed on a flat, clear glass plate. Place the plate and the gel on the copier and cover them with a piece of absorbent paper to prevent the cover of the copier from becoming wet. A photocopy will be quite satisfactory if the resolution of the bands is very good. Photocopies are especially useful with zymograms that are difficult to interpret and require further study. The researcher can conveniently write on the hard copy and create models to aid in the interpretation of the zymogram.

3.15. Gel Storage

Gels may be wrapped in a plastic wrapper to avoid desiccation and then stored in a refrigerator. They may be stored with a small amount of a mixture of 1:1 of glycerol:water in a plastic bag at room temperature.

Chapter 4

ELECTROPHORESIS IN POLYACRYLAMIDE GEL

The polyacrylamide gel electrophoresis (PAGE) technique was introduced by Raymond and Weintraub (1959). Since then, the methodology has been improved and adapted to various applications in a variety of scientific fields. This chapter presents some of the more commonly used PAGE procedures.

4.1. Introduction

Polyacrylamide is a synthetic polymer (Chrambach and Rodbard 1971) prepared by polymerizing the acrylamide monomer into long chains, cross-linked by certain reagents (Hames 1981) into three-dimensional structures (Takacs and Kerese 1984). Separation of proteins in PAGE is accomplished on the basis of both molecular weight and charge (Chrambach and Rodbard 1971). Polyacrylamide gel (PAG) may be prepared so as to provide a wide variety of electrophoretic conditions. The pore size of the gel may be varied to produce different molecular sieving effects for separating proteins of different sizes (Rodbard and Chrambach 1970). PAG also may be prepared at pH values of between 3 and 11 so as to maximize the net charge differences between molecules for better separation (Rodbard and Chrambach 1970). Furthermore, PAGE may be conducted at temperatures from as low as 0°C to room temperature, hence making the technique applicable to the separation of thermolabile molecules such as enzymes (Gabriel and Wang 1969).

4.2. Apparatus for Polyacrylamide Gel Electrophoresis

PAG may be cast in one of two basic shapes, rod (cylindrical) and slab (flat). The dimensions of these shapes are variable. The slab gel whose thickness may range from 0.75 to 1.5 mm is usually preferred because it provides an identical medium for all samples, thus permitting easy photography and densitometry and more efficient cooling during electrophoresis. PAGE may be conducted with the gel in either a vertical or horizontal position; each type requires a different apparatus. Most of the examples given in this chapter are for vertical PAGE. Two models of apparatus for vertical gels are shown in Figures 4-1 to 4-5. The Studier type (Studier 1973) is easy to construct as a homemade unit but lacks the facility for supplemental cooling when operated at high voltage at room temperature. The Hoeffer system allows multiple gels (up to 12, depending on the model) to be electrophoresed simultaneously and also has a system for water cooling when needed.

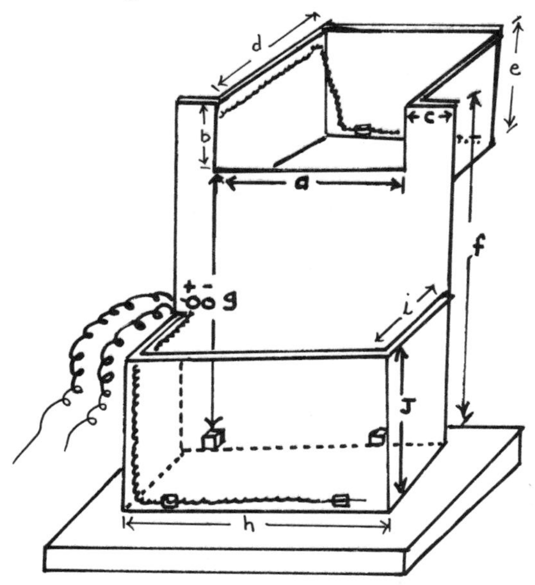

Figure 4-1. A Studier type apparatus for polyacrylamide gel electrophoresis. The dimensions are: $a = 13$ cm, $b = 3$ cm, $c = 2$ cm, $d = 4$ cm, $e = 6$ cm, $f = 16$ cm, $g = 12$ cm, $h = 17$ cm, $i = 4$ cm, $j = 5$ cm.

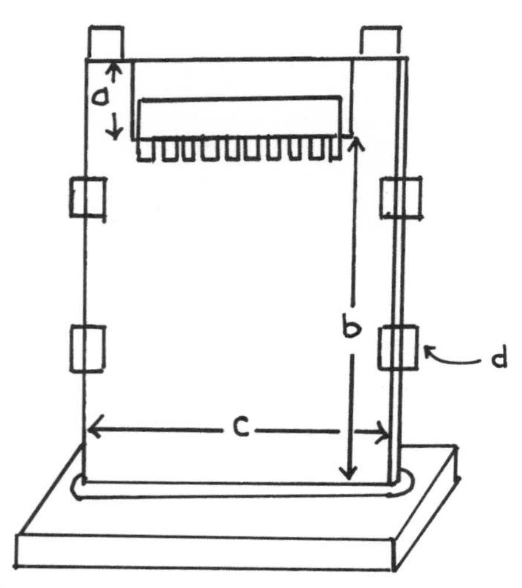

Figure 4-2. A gel casting unit for a Studier type polyacrylamide gel electrophoresis unit. The dimensions are: $a = 3$ cm, $b = 12$ cm, $c = 16$ cm; d = paper clamp.

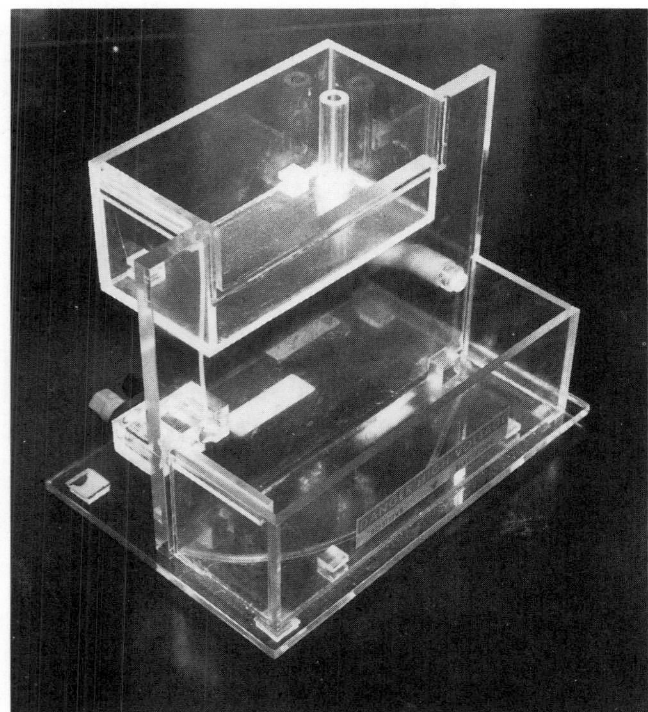

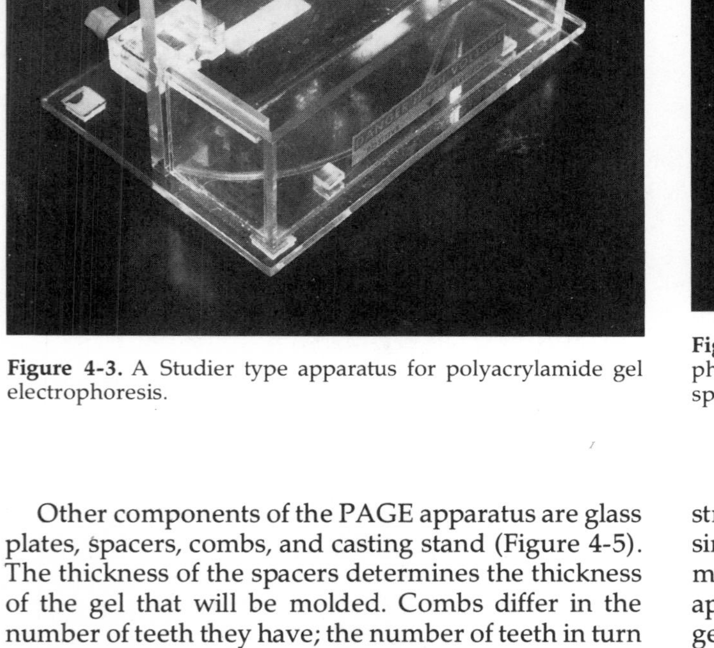

Figure 4-3. A Studier type apparatus for polyacrylamide gel electrophoresis.

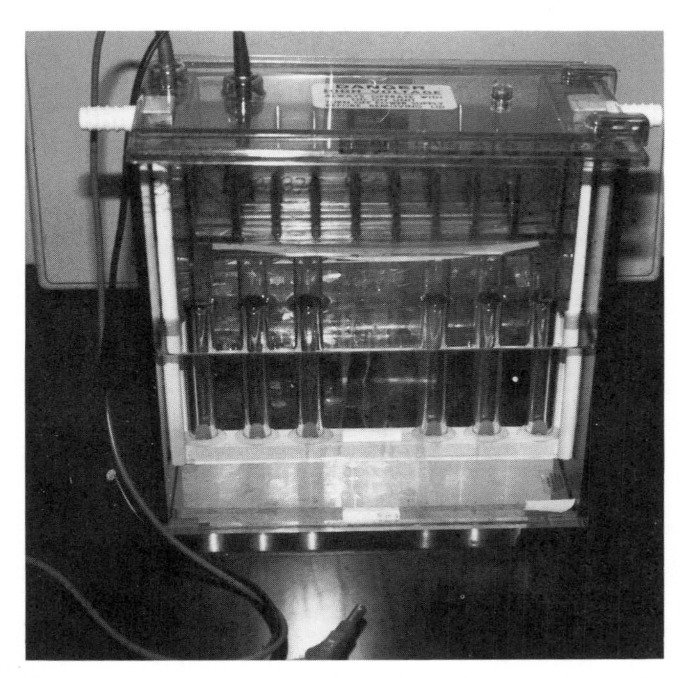

Figure 4-4. A Hoeffer unit for polyacrylamide gel electrophoresis.

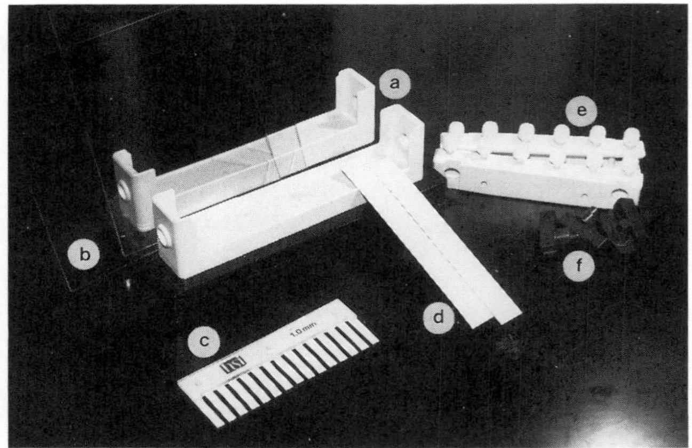

Figure 4-5. Accessories for a Hoeffer polyacrylamide gel electrophoresis unit: (*a*) casting stand, (*b*) glass plates, (*c*) comb, (*d*) spacers, (*e*) clamps, (*f*) cams.

Other components of the PAGE apparatus are glass plates, spacers, combs, and casting stand (Figure 4-5). The thickness of the spacers determines the thickness of the gel that will be molded. Combs differ in the number of teeth they have; the number of teeth in turn determines the number of sample wells that can be obtained.

Following the dimensions provided, one may construct a homemade Studier type PAGE apparatus similar to that shown in Figure 4-3 with the same materials as described above for constructing SGE apparatus (Chapter 3). PAG may be cast like a starch gel for slicing by using a Type I mold for SGE. For horizontal PAG, a lid should be provided for the mold to reduce the contact with oxygen from the air. (Oxygen hinders polymerization of acrylamide.)

4.3. Casting a Gel

Unlike starch gels, polyacrylamide gels require no heating in their preparation. The major difference between starch and polyacrylamide, however, is the caution required in their preparation, since unpolymerized acrylamide is a neurotoxin.

4.3.1. Materials

The materials required to cast a polyacrylamide gel depend on the type of electrophoretic procedure used and include the following: (*a*) PAGE unit, (*b*) filter flask, (*c*) power pack, (*d*) beakers, (*e*) acrylamide and bis-acrylamide, (*f*) measuring cylinders, (*g*) catalysts, (*h*) pipette, (*i*) buffers.

4.3.2. Assembling the Casting Unit

The plates used in a casting unit must be very clean and smooth. They should be washed and rinsed with distilled water and then wiped with a tissue soaked in ethanol. Several commercial cleaning solutions are also available for this purpose. The manufacturer's directions must be followed in assembling the casting unit. The examples in this chapter are based on the Hoeffer system.

Generally, two spacers are sandwiched between two plates, on the opposite sides of a "spacer mate," and clamped together (Figure 4-6). It is critical that the edges of plates and spacers be flush with each other (no overlapping), or else the gasket in the base of the casting stand will fail to create an effective seal, thus leading to leaking of the gel mixture when poured between the clamped plates. A comb is inserted at the top end of the space between the plates to cast the wells that will contain the protein samples. The casting apparatus must be set up before the gel mixture is prepared. Gel mixtures start to polymerize soon after the addition of a catalyst. Furthermore, the apparatus must be set on a plain surface before the gel is cast to avoid a slanted surface upon polymerization. Some commercial PAGE units have a built-in liquid level to help in setting up the casting apparatus.

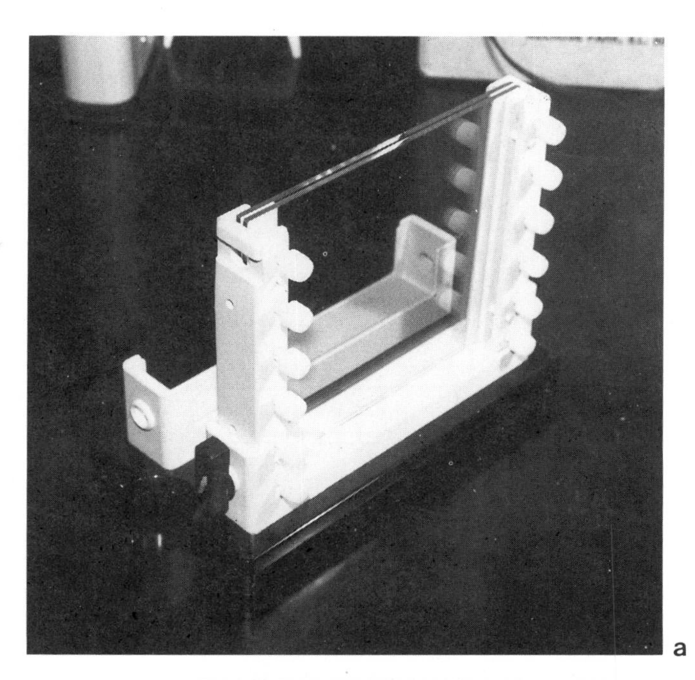

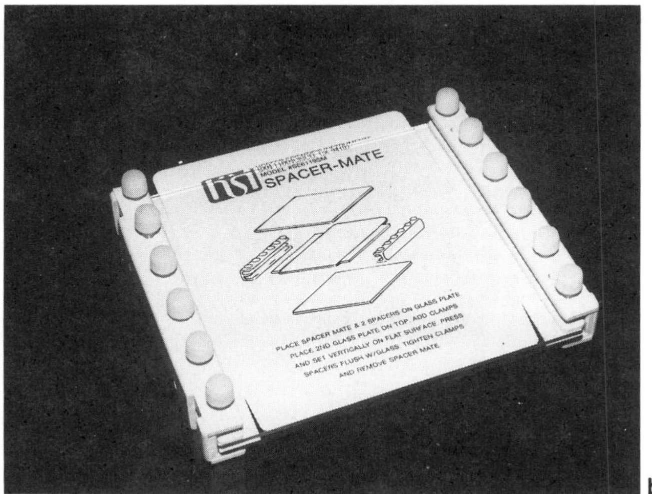

Figure 4-6. Casting a gel for a Hoeffer unit for polyacrylamide gel electrophoresis: (*a*) glass plates and spacers clamped together with the aid of a spacer-mate, (*b*) clamped plates installed in a casting stand.

4.3.3. Preparing the Gel Mixture

A PAG has two critical properties, pH and pore size, both of which are within the control of the investigator. As mentioned above, acrylamide is a monomer that has to be polymerized before use. Three major ingredients are used in PAG preparation: (*a*) acrylamide, (*b*) cross-linker (*N, N*′-methylene-bis-acrylamide, simply called bis or bis-acrylamide), and (*c*) polymerization initiator or catalyst [e.g. ammonium persulfate, riboflavin, *N,N,N′N′*-tetramethylenediamine (TEMED) and light (day light flourescent].

For reproducibility of results, either the acrylamide and bis should be recrystalized (Loening 1967) or high purity reagents should be purchased. The proportions of the three gel ingredients listed above determine the rate of polymerization and the pore size of a PAG. The total acrylamide concentration is designated %*T*; the concentration of the cross-linking agent is designated %*C* (Hjerten 1962; Fawcett and Morris 1966). These parameters are calculated as follows:

$$\%T = (\text{g acrylamide} + \text{g bis-acrylamide})/(\text{g acrylamide}) \times 100.$$

$$\%C = (\text{g bis-acrylamide})/(\text{g acrylamide} + \text{g bis-acrylamide}) \times 100.$$

where g is gram.

As the %*T* increases, the pore size of the gel decreases. The minimum pore size is attained at $C_{bis} = 5\%$ (Hames 1981). The operational %*T* ranges between 3 and 30% w/v whereas that for %*C* is between 1 and 25% of the total monomer (Chrambach and Rodbard

1971). Polymerization may be accomplished chemically or photochemically. The two most commonly used polymerization systems are ammonium persulfate-TEMED and riboflavin-TEMED; the latter requires light for initiation (Hames 1981).

Because a polymerized gel stores poorly, polymerization must be carried out prior to each electrophoresis (Chrambach and Rodbard 1971).

REMINDER

1. Acrylamide and bis-acrylamide are neurotoxins, so wear gloves when handling them.
2. Add catalysts to the gel mixture as the last step.
3. After adding the catalysts, pour the gel mixture immediately before polymerization begins. Pouring may be accomplished by using a special pump, a syringe, or pipette.
4. Avoid trapping air bubbles in the gel.
5. Oxygen is a major polymerization inhibitor and may be excluded by deaeration of the gel mixture or by saturation with an inert gas (Hjerten et al. 1969). Inhibition of polymerization by oxygen is especially a problem during acid PAGE and at low (0°C) temperatures (Hjerten et al. 1969).
6. Polymerization should be conducted at temperatures similar to those for electrophoresis to avoid thermal expansion or contraction of the gel (Chrambach and Rodbard 1971). The temperature should be constant throughout polymerization in order to achieve uniformity of the gel. Keep the solutions for polymerizing gels in a refrigerator (4°C).
7. The duration of polymerization may be as little as 5–15 min, but a gel is usually left to sit unperturbed for about 30 min (Kingsbury and Masters 1970). Acid gels usually require a longer time for polymerization than basic ones.
8. All the PAG mixture does not polymerize. A polymerization efficiency of a least 95% is desirable, and 98% is attainable (Chrambach and Rodbard 1971).
9. To eliminate the meniscus at the top of the gel, a thin layer of water is sometimes floated on top of the polymerizing gel. Pour the water layer off after polymerization. The water layer also serves as a barrier to block out oxygen, which inhibits polymerization (see above).
10. Carefully remove the comb after polymerization. If the walls of the wells bend in the process, reposition them with a fine tool such as the needle of a syringe. Rinse the wells with the running buffer and fill them with same prior to loading the samples.
11. When a multiphasic gel is being cast, the stacking gel usually polymerizes sooner than the resolving gel and must be poured quickly.

4.4. Sample Preparation

The method of sample preparation chosen for a study depends on the type of PAGE buffer system to be used.

4.4.1. Nondissociating Buffer System

Samples are prepared for a nondissociating buffer system in a way that will ensure that native proteins do not become denatured. The general procedure is as follows:

1. Grind the samples (where applicable).
2. Incubate the sample in the extraction buffer.
3. Centrifuge.
4. Mix supernatant with "anchor" solution in an appropriate ratio (e.g. 1:1). The "anchor" solution is a high-density solution (e.g. glucose, glycerol) whose function is to keep the sample at the bottom of the well and prevent diffusion into the running buffer. It usually contains a tracking or marker dye.

4.4.2. Dissociating Buffer System

Surfactants may be used in PAGE for special applications. The commonly used surfactants are urea and sodium dodecyl sulfate (SDS). Dissociation is achieved by heating the protein sample in an anionic detergent in the presence of a reducing agent. The surfactants bind the polymers of the proteins and in the process confer upon them negative charges proportional to their lengths (Shapiro et al. 1967; Weber and Osborne 1969). In other words, the resultant SDS-polymer complexes have equal charge densities (charge per unit length). The acquired negative charges overwhelm the original charges possessed by the polypeptide molecules. Further, the SDS-polypeptide complexes migrate in the appropriate medium at rates proportional to the sizes of the polypeptides. This property yields information on both the polypeptide composition and the molecular weight of the constituent polypeptides when samples are electrophoresed along with molecular weight markers in the same gel (Hames 1981). SDS systems have been developed by Ornstein (1964), Davis (1964), and Laemmli (1970) for discontinuous buffers. Weber and Osborne (1969) developed a continuous buffer system.

The general steps in preparing samples for dissociating PAGE systems are as follows:

1. Heat sample in extraction buffer to boiling for 1–5 minutes.
2. Centrifuge.
3. Add "anchor" and tracking dye (some recipes include these in the extraction buffer).

4.4.3. Preparing Samples from Very Dilute Source Material

Sometimes samples are too dilute to be used for electrophoresis. The protein has to be concentrated prior to use by employing one of various methods, including lyophilization and ammonium sulfate precipitation (Hames 1981). For example (Hames 1981):

1. Mix sample with five volumes of cold acetone.
2. Incubate at −20°C for 10 min.
3. Centrifuge at 10,000 g for 5 min.
4. Wash repeatedly with 1:1 v/v of ethanol:ether.
5. Dissolve protein pellet in appropriate buffer for use.

4.5. Loading Samples

As mentioned above, high-density solutions are mixed with samples to anchor the samples to the bottom of the wells. The marker or tracking dye enables the researcher to follow the progress of electrophoresis and also provides a reference point for the calculation of electrophoretic mobilities of the various molecules. The choice of a tracking dye depends on the pH of the electrophoretic conditions. For acid PAGE, methyl green is commonly used whereas bromophenol blue is preferred for basic PAGE. The amount of samples are loaded in each sample well in amounts according to the concentration of the protein in the sample. A loading volume of 10 to 30 μl is adequate for slab gels (Hames 1981). Overloading of samples may result in poor resolution and distortion of bands and in unwanted interaction between adjacent sample lanes. On the other hand, insufficient amounts of sample may result in faint or undetectable bands.

In the Hoeffer system, the top buffer tray is attached to the glass plates (while in the casting stand) and the tray partially filled with resolving buffer. If only one gel is to be electrophoresed, a "dummy gel" (a thick plastic slab supplied by the manufacturer) should be installed. Samples are loaded into the wells via a Hamilton syringe. After loading, the lower set of cams are removed and the plates with the upper buffer tank attached are lifted out of the casting stand and installed into the lower buffer tank. Additional running buffer should be added to the upper tank to cover up the electrode. Nondissociating PAGE requires cooling (4 to 8°C) to preserve the physiological activity of the proteins in their native form. The lower tank should have enough buffer to cover the plates. A water-cooling system should be installed or else the electrophoresis conducted in a cold room. For SDS-PAGE, filling the lower tank or providing supplemental cooling is not necessary. The buffer should cover about 2 to 4 cm of the bottom of the plates.

Electrical connections should then be made to the appropriate electrodes, bearing in mind that certain situations require a reversal of polarity. The running current and voltage will depend on the gel concentration and the dimensions of the gel. Generally, SDS-PAGE requires longer duration and is conducted at lower current than conventional PAGE (2–5 mA overnight vs 25–30 mA for 4 to 6 hr).

4.6. Staining a Gel

After electrophoresis is completed, turn off the power to begin unloading the gel for staining. Remove the cams and spacers so the plates can be pried open (use a nonmetallic wedge to avoid chipping the glass) and the gel removed. For multiphasic gel, the stacking gel is cut off before lifting up the remainder of the gel or pushing it to slide off the plate into the staining dish. As in SGE, the top corner of the gel above the first sample lane may be cut off to indicate the beginning of loading of the samples for the purpose of scoring. Gels of low acrylamide concentration should be handled with extra care to avoid tearing. The gel is placed in a staining tray and submerged in a dye solution. The most commonly used dye for staining general proteins is Coomassie brilliant blue R-250 (Chrambach et al. 1967). Other dyes include Amido black (Davis 1964) and fast green (Gorovsky et al. 1970). Staining requires at least several hours to complete and routinely is carried out overnight.

4.7. Destaining a Gel

Staining usually produces intensely colored bands over a dark background. For better visualization of gels, staining is normally followed by destaining to remove excess stain and to clear the dark background.

4.8. Post-staining Handling of a Gel

Gels may be photographed immediately or stored for later viewing. Gels may be stored wet or dry. Commercial dryers are available for drying gels. Alternatively, a gel may be stored in a sealed plastic bag containing 7% acetic acid (Hames 1981).

4.9. Selected Protocols for Polyacrylamide Gel Electrophoresis

PAG is prepared readily from stock solutions. Some of the commonly used stocks are presented below.

1. Acrylamide-bis-acrylamide stock

WORKING RECIPE:

Acrylamide = 30 g
Bis-acrylamide = 0.8 g
Water = 100 ml

STORAGE: light-sensitive; store in dark bottle in refrigerator.
STABILITY: Use within 4–6 wk.
CAUTION: Neurotoxin; avoid inhalation or contact with skin.
OTHER NOTES: Filter the solution through Whatman No. 1 paper after preparation. Unless very high-purity grade is used, both reagents may require purification (e.g. recrystallization).

2. 10% Ammonium persulfate

WORKING RECIPE:

Prepare on w/v basis as follows:
Ammonium persulfate = 1.0 g
Water = 10 ml

STORAGE: Do not store; prepare fresh when needed.

3. Riboflavin

WORKING RECIPE:

Prepare on w/v basis as follows:
Riboflavin = 0.004 g
Water = 100 ml

STORAGE: Light sensitive; store in dark bottle in a refrigerator.
STABILITY: Stable for 4–6 wk.

4. SDS

WORKING RECIPE:

Prepare on w/v basis as follows:
SDS = 10 g
Water = 100 ml

STORAGE: Room temperature.
STABILITY: Use within 4–6 wk.
OTHER NOTES: Dissolve SDS completely to produce a clear solution.

Takacs and Kerese (1984) have listed some of the commonly used buffer recipes in PAGE. The recipes presented below have been selected to meet a variety of needs and include examples for acid and basic PAGE. The protocols have been coded for quick reference. Buffer stocks are often prepared such that dilution is required before use. The relative proportions of the components of each recipe have been given so that the researcher can prepare any amount of a particular buffer by simply changing the proportions of each component. With some understanding of the basis of PAGE, one may be able to modify the amounts of the components of a buffer for specific reasons. Except otherwise indicated, the selected recipes are be based on the following stock solutions:

1. Acrylamide-bis-acrylamide

Acrylamide	10 g	30 g	60 g
Bis acrylamide	2.5 g	0.8 g	1.4 g
Code	A1	A2	A3

2. Electrode buffers

		Buffer solution components			
Code	pH	tris (g)	Glycine (g)	β-Alanine (ml)	AcOH
E1	8.3	6	28.8	—	—
E2	4.5	—	—	31.2	8.0
E3	4.0	—	28.1	—	3.06

All the quantities of ingredients are for preparing 100 ml of 10× buffer solution. Dilute ten times with distilled water before use. AcOH = acetic acid.

3. Gel buffers

		Buffer solution components					
Code	pH	tris (ml)	TEMED (ml)	1 M HCl (ml)	1 N H_3PO_4 (ml)	AcOH (ml)	1 M KOH (ml)
G1	8.9	36.6	0.23	48	—	—	—
G2	4.3	—	4.0	—	—	17.3	48
G3	2.9	1.15	—	—	53.25	12	
G4	6.7	—	0.46	—	—	2.87	48
G5	6.9	5.7	—	—	25.6	—	—
G6	5.9	—	—	—	—	2.95	48

All the quantities of ingredients are for preparing 100 ml of 10× buffer solution. Dilute ten times with distilled water before use. AcOH = acetic acid.

4. Ammonium persulfate (AP)

Code	Water (ml)	AP (g)
P1	100	0.14
P2	100	0.28
P3	100	2.8
P4	100	10.0

5. Riboflavin solution

Code	Water (ml)	Riboflavin (g)
R1	100	0.004

6. Sucrose solution

Code	Water (ml)	Sucrose (g)
S1	100	40

7. Tracking dye solution:

This could be 0.001% bromophenol blue or methyl green, depending on the pH of the electrophoretic conditions.

4.9.1. Selected Recipes for Native Basic PAGE

At a basic pH, electrophoresis is conducted such that the protein molecules migrate anodally. The catalyst should be the last item to add to the gel mixture in each recipe. Except otherwise indicated, the codes refer to those mentioned in the stock solutions 1 to 7 above.

Example 1

RESOLVING GEL: 3.75% pH = 8.9

Code	Proportion
A2	1
G1	1
R1	1
Water	5

STACKING GEL: 3.12% pH = 6.9

Code	Proportion
A1	2
G5	1
R1	1
S1	4

RUNNING BUFFER (TANK): E1, pH 8.3

Example 2

RESOLVING GEL: 7.5% pH = 8.9

Code	Proportion
A2	2
G1	1
P1	4
Water	1

STACKING GEL: 3.12% pH = 6.9

Code	Proportion
A1	2
G5	1
R1	1
S1	4

RUNNING BUFFER: E1, pH 8.3

Example 3

RESOLVING GEL: 15% pH = 8.9

Code	Proportion
A3	2
G1	1
P1	4
Water	1

STACKING GEL: 3.12% pH = 6.9

Code	Proportion
A1	2
G5	1
R1	1
S1	4

RUNNING BUFFER: E1, pH 8.3

4.9.2. Selected Recipes for Native Acid PAGE

A note of caution to observe when conducting an acid PAGE is that the direction of electrophoresis is reversed from that of basic PAGE so that proteins migrate cathodally. Failure to reverse the terminals will result in samples migrating into the buffer tank and being lost. As mentioned above, an acid gel usually requires an extended period of polymerization.

Example 1

RESOLVING GEL: 7.5% pH = 4.3

Code	Proportion
A2	2
G2	1
P2	4
Water	1

STACKING GEL: 3.12% pH = 6.7

Code	Proportion
A1	2
G4	1
R1	1
Water	4

RUNNING BUFFER: E2, pH 4.5

Example 2

RESOLVING GEL: 15% pH = 4.3

Code	Proportion
A3	2
G2	1
P2	4
Water	1

STACKING GEL: 3.12% pH = 6.7

Code	Proportion
A1	2
G4	1
R1	1
Water	4

RUNNING BUFFER: E2, pH 4.5

Example 3

RESOLVING GEL: 7.5% pH = 2.9

Code	Proportion
A2	2
G3	4
P3	2

STACKING GEL: 3.12% pH = 5.9

Code	Proportion
A1	2
G6	1
R1	1
S1	4

RUNNING BUFFER: E3, pH 4.0

4.9.3. Further Comments on Acid PAGE

Since polymerization at acidic pH is a very slow process, procedures have been developed to speed up the process. For example, Jordan and Raymond (1969) developed the following catalyst for rapid polymerization:

Ascorbic acid	0.1%
Ferrous sulfate	0.0025%
Hydrogen peroxide	0.03% (add just before pouring)

Polymerization temperature = 5–10°C
Duration of polymerization = 3–5 min

4.9.4. Recipes for SDS-PAGE
SDS-PAGE may be conducted in either a discontinuous or continuous buffer system. Either way, the method of sample preparation denatures and dissociates proteins into component polypeptide chains.

4.9.4.1. Extraction Buffers and Sample Preparation The examples given below are used along with recipes from Laemmli (1970) and Weber and Osborne (1969) presented further on in this chapter.

1. BUFFER: 0.0625 M tris-HCl, pH 6.8 (Laemmli 1970).
2% w/v SDS
10% v/v Glycerol
5% v/v β-Mercaptoethanol
0.001% w/v Bromophenol blue
(w/v is weight for volume and v/v is volume for volume)

Boil sample in the extraction buffer for 1–5 min and centrifuge (10,000–20,000 g for 15–20 min). Load 5–50 μl of supernatant.

2. BUFFER: 0.01 M Sodium phosphate, pH 7.0 (Weber and Osborne 1969).
1% w/v SDS
1% v/v β-Mercaptoethanol

Incubate sample in buffer for 2 hr at 37°C. Mix supernatant with the following mixture in a ratio of 1:1

0.05% Bromophenol blue
in H_2O = 3 μl
Glycerol = 1 drop
β-Mercaptoethanol = 5 μl
Extraction buffer = 50 μl
Load 5–50 μl of sample.

4.9.4.2. Discontinuous Buffer System (Laemmli 1970) Some of the stock solutions in this example have been prepared to 4× of the original concentration and must therefore be diluted before use.

1. Acrylamide stock.
Acrylamide:bis-acrylamide = 30:0.8 by weight
Acrylamide = 30 g
Bis-acrylamide = 0.8 g
Water = 100 ml
2. Resolving gel buffer (4×), pH 8.8.
1.5 M tris-HCl
0.4% w/v SDS
(Dilute 4× before use)
3. Stacking gel buffer (4×), pH 6.8.
0.5 M tris-HCl
0.4% w/v SDS
(Dilute 4× before use)
4. 10% Ammonium persulfate solution (AP)
100 mg per 1 ml of water
5. Running buffer (4×), pH 8.3.
0.1 M tris
0.768 M Glycine
Dilute to 1× then add SDS to final concentration of 0.1% w/v.
6. Preparing a 30 ml gel.

	Resolving gel (15%)	Stacking gel (6%)
Water	7.4 ml	5.4 ml
Buffer	7.5 ml	2.5 ml
Acrylamide	15.0 ml	2.0 ml
TEMED	10.0 μl	5.0 μl
10% AP	100.0 μl	75.0 μl

TRACKING DYE: Bromophenol blue
POWER: 3 mA for 7 hr

4.9.4.3. Continuous Buffer System (Weber and Osborne 1969)

1. Resolving gel buffer, pH 7.1.
$NaH_2PO_4.H_2O$ = 7.8 g
$Na_2HPO^4.7H_2O$ = 38.6 g
SDS = 2 g
Water = 1000 ml
2. Acrylamide.
Acrylamide = 22.2 g
Bis-acrylamide = 0.6 g
3. Ammonium persulfate (AP) = 15 mg in 1 ml water.
4. Running buffer.
Dilute gel buffer with water in a ratio of 1:1.
5. For a 10% gel.
Acrylamide = 13.5 ml
Gel buffer = 15 ml
AP = 1.5 ml
TEMED = 45 μl
POWER: 8 mA for 4 hr.

4.10. Determinating the Molecular Weight of Proteins from SDS-PAGE (Weber and Osborne 1967)

A PAG swells about 15% due to the acetic acid in staining and destaining solutions. The calculation of molecular weight should therefore include data on the gel length before and after staining, in addition to the mobilities of the protein molecules and the tracking dye. The relative mobility (R_f) is calculated as

$$R_f = \frac{\text{distance of protein migration}}{\text{gel length after staining}} \times \frac{\text{gel length before staining}}{\text{distance of tracking dye migration.}}$$

The R_f values are then plotted against the known molecular weights (standards that were electrophoresed along with the samples), expressed on a semi-logarithmic scale, to obtain a linear correlation (Figure 4-7). The R_f values of the unknown molecular weights of the samples are read off the curve to find their corresponding molecular weights.

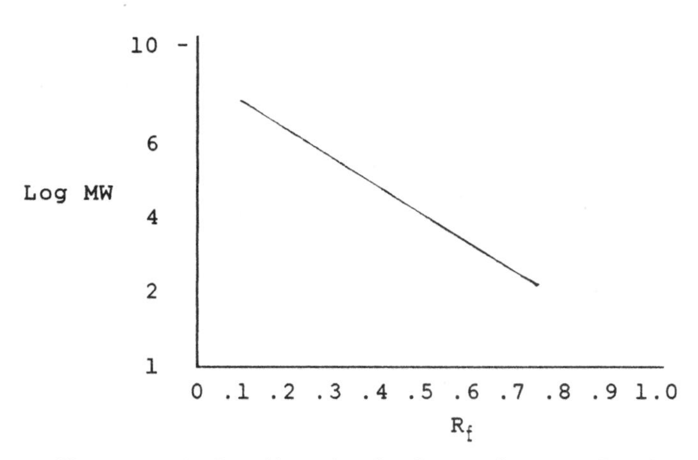

Figure 4-7. A plot of log of molecular weight versus R_f values.

4.11. Preparing Gels for Visualization

Visualization in PAGE is accomplished by staining a gel in an appropriate dye solution. The staining step sometimes must be preceded by a fixing step. Some of the frequently used methods are described below.

4.11.1. Method 1 (Meyer and Lambert 1965)

Fix in 5:5:1 solution of methanol:acetic acid:water.
Stain in fixing solution containing 0.25% Coomassie BB R-250.
Destain in 3.75% acetic acid.

4.11.2. Method 2

Fix in 7% acetic acid for 1 hr.
Stain in fixing solution containing 1% amido black 10 B for 12 hr.
Destain in 7% acetic acid.

4.11.3. Method 3 (Gorovsky et al. 1970)

Fix in 7% acetic acid
Stain in fixing solution containing 1% fast green, at 20°C for 2 hr.
Destain in 7% acetic acid.

4.11.4. Method 4 (Chrambach 1966)

Fix in 12.5% trichloacetic acid (TCA) for 1 hr.
Stain in 1% Coomassie BB R-250 diluted 20 times with fixing buffer.
Destain in 10% TCA first, decant and add 7% acetic acid.

4.11.5. Method 5 (Khan and Rubin 1975)

Stain in 1% Coomassie BB R-250 in 25% TCA for 2 hr.
Destain in 8% acetic acid for 2–3 days.

Depending on the type of procedure used, gels from PAGE may be subjected to histochemical staining as is done in starch gel electrophoresis. The gel may be submerged in a staining solution prepared for a specific enzyme. Alternatively, the method of agar overlay may be applied.

4.12. Further Comments on PAGE

A successful PAGE depends largely on the judicious choice of buffers. Knowledge of the samples' molecular weight, charge, pH, and tendency to dissociate is advantageous (Takacs and Kerese 1984). Some of this information is supplied with the recipes provided here.

Ammonium persulfate is one of the most preferred catalysts of the polymerization reaction in PAGE. Although this reagent can be obtained in high purity, its strong oxidation effects may produce staining artifacts with some proteins, and possibly inactivate others (Takacs and Kerese 1984). The addition of antioxidants such as dithiothreitol (DTT) and 2-mercaptoethanol to the polymerization mixture can reduce or eliminate the above problem (King 1970).

4.13. Selected Advanced Techniques of Protein Separation in Polyacrylamide Gel

High-resolution techniques are sometimes employed in protein separation. This section introduces two that are commonly used.

4.13.1. Introduction to Isoelectric Focusing

Isoelectric focusing (IEF) is a high-resolution technique for separating proteins and other amphoteric molecules. It is particularly effective for testing homogeneity of protein samples but is quite expensive to employ. Kolin (1953), Svensson (1961), and Vesterburg and Svensson (1966) are credited with developing the principles of IEF and its practical application as described below.

A critical difference between IEF and conventional electrophoresis is that whereas conventional electrophoresis is conducted at a constant buffer pH, IEF effects separation in a uniform pH gradient. IEF is a two-stage technique. First, a pH gradient is produced by introducing carrier ampholytes (synthetic aliphatic compounds with different charges) into a support medium. When placed in an electric field, the current causes the ampholytes to be arranged in the medium according to their isoelectric points (pI) (point of zero net charge). This creates a pH gradient in which the anodal region of the medium is occupied by the most acidic ampholytes while the most basic ones migrate to the cathodal region. In the second stage, protein samples (which may be loaded in a variety of ways) migrate along the pH gradient in the support matrix until they reach their respective pIs. At those locations, the proteins completely lose electrophoretic mobility due to their zero net charge. The effect of these events is that proteins of identical pI become concentrated in a very narrow band. Should a molecule stray away from its pI location, it immediately acquires a charge and becomes electrophoretically mobile once again, whereupon it must again migrate back to its pI. It is this anti-diffusion property of IEF that is responsible for the high resolution of bands in focused gels.

To use the technique effectively, certain aspects of its operation must be understood, the most important of which are as follows.

4.13.1.1. Properties and Types of IEF Support Media A suitable medium for IEF should be nonrestrictive and have anticonvective properties. The pore size should be such that migrating molecules are not impeded. The medium should have also low endoosmotic effect. The most suitable media that meet the above criteria are acrylamide and sephadex (Righetti and Drysdale 1974). A very low gel concentration (as for a stacking gel) or a mixture of agarose and acrylamide provides maximal mechanical stability of gel and minimal molecular sieving (Chrambach and Rodbard 1971).

Focusing may be done in various formats: vertical slab, horizontal slab, or cylinders. Slab units allow casting of very thin gels for efficient heat dissipation and ready comparison of similar samples, whereas cylinder units permit the analysis of dilute samples and more ready adaptation of IEF products to other analytical methods such as densitometry and two-dimensional PAGE (Righetti and Drysdale 1974). Commercially precast gels may also be purchased for IEF.

4.13.1.2. Choice of Ampholyte Range The pH gradient used must be stable to yield reproducible results of IEF that are amenable to precise interpretation. Although a wide pH range will give a higher probability that all the constituent proteins in a sample are revealed, a narrow pH range of ampholyte pIs gives the best resolution of bands (Chrambach and Rodbard 1971). A concentration of 4% (w/v) of ampholyte produced a linear pH gradient whereas a gel with 1% ampholyte produced irregular results (Righetti and Drysdale 1974).

4.13.1.3. Choice of Electrolyte A critical consideration in the choice of electrolyte is that the pH of anolyte and catholyte encompasses the pH gradient to be developed in the gel (Righetti and Drysdale 1974).

4.13.1.4. Loading Samples for IEF Samples may be applied in a variety of ways, depending on the apparatus. Samples are usually loaded on polymerized gels at the cathodal end in a solution of ampholyte and a dense liquid such as 10% glycerol. Applications may be "acute" (single dose) or "chronic" (multiple consecutive doses of dilute samples). In slab gels, samples may be applied by layering them on top of the gel in parallel tracks. Leaback and Rutter (1968) applied samples to indentations in the gel. The maximum amount of protein that can be focused as a distinct sharp band depends on its proximity to other focused bands. If the bands of the samples are focused close together, the amount of protein that is loaded becomes more critical. Overloading can cause the minor sample components to be focused adequately for visualization.

4.13.1.5. Duration of Electrolysis The choice of wattage for IEF depends on the type of apparatus and the temperature at which electrolysis is conducted. Sharper resolutions are obtained with rapid separation at a high voltage (Svensson 1961). If efficient cooling is provided, focusing can be achieved in about one hour at 1000 volts and 50 Watts, but there is a danger of pH gradient instability (Wadstrom 1974). The duration also depends on the concentration of the ampholyte and the molecular weight of the proteins. Larger molecules require a longer period of electrofocusing (Righetti and Drysdale 1974).

4.13.1.6. Staining of Gels One of the dyes most commonly used in direct staining is Coomassie brilliant blue. For example (Diezel et al. 1972):

1. Soak gel in 12.5% TCA (40 times excess) for 5 min.
2. Add 0.25% aqueous Coomassie BB G–250 to a final concentration of 0.0125%.
3. Stain for 30 min.
4. Transfer into 5% acetic acid.

4.13.1.7. Determination of Realized pH Gradient Determining the actual pH gradient may be necessary to estimate the pI of the protein constituents of interest. A quick method is to include standard proteins of known pI in the same gel and assign pI by comparison of the sample to the standard. Alternatively, one can run two gels (or use just one) under identical conditions. The slab gel is first sliced into strips according to the number of samples. Each strip is then further sliced into short pieces (one may use a ruler to guide). Each section is soaked in an eluate of distilled water or dilute salt solution, e.g. 0.025 M KCl for 30 min (An der Lan and Chrambach 1981). The reading of the pH requires the use of a microelectrode. Devices for measuring pH gradient are available for purchase (An der Lan and Chrambach 1981).

4.13.2. An Isoelectric Focusing Protocol (O'Farrell 1975)

The IEF procedure described below was designed for use with cylinder gel units. Apart from its popularity, I selected it because it is part of another protein separation procedure (2-D electrophoresis) also presented below. The stock solutions required for the IEF procedure are as follows:

4.13.2.1. Stock Solutions For quick reference, the stock solutions required for the IEF procedure have been coded by capital letters in parentheses.

Lysis buffer (A):
9.5 M urea
2% Ampholines (consisting of 1.6% pH 5–7 plus 0.4% pH 3–10)

5% β-Mercaptoethanol
Store aliquots in freezer.

Sonification buffer (B):
0.01 M tris-HCl, pH 7.4
5 mM $MgCl_2$
50 μg/ml pancreatic RNase

Pancreatic DNase (C):
1 mg/ml in 0.01 M tris-HCl, pH 7.4
1 mM $MgCl_2$
Store aliquots in freezer.

Acrylamide for IEF gel (D):
28.38% w/v Acrylamide
1.72% bis-Acrylamide
Store in the dark in refrigerator.

Nonidet P–40 solution (E):
10% w/v NP–40 in water

Ampholines (F):
Use as supplied.
40% w/v solutions

Ammonium persulfate (G):
10% solution (prepare fresh every 2 wk)

Gel overlay solution (H):
8 M urea (store aliquots in freezer)

Anode electrode solution (I):
0.01 M H_3PO_4

Cathode electrode solution (J):
0.02 M NaOH (degas and store under vacuum)

Sample overlay solution (K):
9 M urea
1% Ampholines (consisting of 0.8% pH 5–7 plus 0.2% pH 3–10)
Store in freezer.

Lower gel buffer (L):
1.5 M tris-HCl, pH 8.8
0.4% SDS

Upper gel buffer (M):
0.5 M tris-HCl, pH 6.8
0.4% SDS

4.13.2.2. Preparing a Gel for Isoelectric Focusing The example given below is for preparing 10 ml of gel for use in an IEF unit that will hold 0.5 ml of gel mixture per tube. The steps are as follows: To a side arm flask (125 ml) add 5.5 g urea, 1.33 ml (D), 2.0 ml (E), 1.97 ml water, 0.4 ml ampholines (pH 5–7), 0.1 ml ampholines (pH 3–10). Swirl flask to dissolve the urea completely then add 10 μl 10% ammonium persulfate. Degas for 1 min and add 7 μl TEMED.

Load the tubes (sealed at bottom with Parafilm™) to about 5 mm from top. Overlay with solution H for 1–2 hr. Remove solution H and replace it with 20 μl of A and overlay with small amount of water. Allow the gel to set for 1–2 hr. Remove the Parafilm™ and cover the ends of the tubes with dialysis membrane (hold in place with 3 mm section of latex tubing). Place the whole setup in a standard tube gel electrophoresis chamber. Remove the lysis buffer (A) and the water from the gel surface and add 20 μl fresh buffer A. Fill the tubes with 0.02 M NaOH, the lower reservoir with 0.01 M H_3PO_4, and the upper reservoir with 0.02 M NaOH (degas to remove CO_2).

After loading, pre-run at 200 V for 15 min, then 300 V for 30 min, and finally 400 V for 30 min. Turn off the power and empty the upper reservoir. Remove the lysis buffer and NaOH from gel surfaces. Using a suitable apparatus (e.g. micropipette), load the samples onto the gels and overlay with 10 μl of buffer K followed by the same amount of 0.02 M NaOH.

Refill the chamber and run the gels at 400 V for 15 hr or 800 V for 1 hr. (Note: V × hr should be more than 5,000 Vhr but less than 10,000 Vhr for full-length gels. The minimum final V should be 400. Too high V causes distortion of bands.

4.13.2.3. Post-focusing Treatment of Gels After IEF, the extruded gels may be handled in one of the following ways:

1. Freeze.
2. Equilibrate and then freeze.
3. Use as is for other applications.

4.13.2.4. Protocol for Equilibration Extrude the gels into 5 ml SDS sample buffer containing 10% w/v glycerol, 5% v/v β-mercaptoethanol, 2.3 w/v SDS, and 0.0625 M tris-HCl pH 6.8 at room temperature and place on a shaker for at least 30 min. (For better results, shake for 2 hr.)

4.13.3. Introduction to Two-dimensional Electrophoresis of Proteins

Two-dimensional (2-D) electrophoresis is a powerful separation technique for complex proteins. Proteins are separated in two directions (dimensions) in two runs so that the proteins that comigrated to one location after the first run but are not identical can be separated in the second run by electrophoresing in a direction at a right angle to the first run (Figure 4-8). The method described here, developed by O'Farrell (1975), is one of the most commonly used. Each dimension separates proteins according to an independent parameter to avoid distributing proteins along the diagonal instead of over the entire surface of the gel (O'Farrell 1975). The separation in the first dimension is accomplished on the basis of charge by isoelectric focusing, whereas the separation in the second dimension is by molecular weight. The O'Farrell method combines IEF and SDS-PAGE (Laemmli 1975). 2-D electrophoresis can separate over 1000 proteins, which appear as spots on the gel (O'Farrell 1975). The analysis of 2-D gels is rather complex, and requires assigning x–y coordinates to the bands.

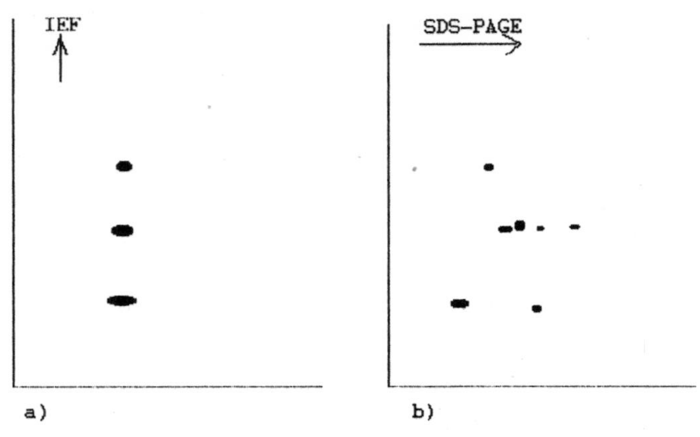

Figure 4-8. Protein separation by two-dimesional electrophoresis: (*a*) separation in the first direction, (*b*) separation in the second direction, showing additional spots.

The two component methodologies of 2-D electrophoresis are described above. What remains to explain is how to load the IEF gel onto a slab gel for the second-dimension run. The method used will depend on the particular apparatus. Generally, however, the extruded IEF cylindrical gel is positioned on top of the opening of the slab gel assembly and held in place by pouring 1% agarose gel onto it. After waiting about 5 min for the agarose to set, the casting stand is removed and the slab placed in the lower tank for normal SDS-PAGE.

4.13.4. Densitometry

Densitometry is a technique for quantitative analysis of proteins after electrophoresis. The rationale of the procedure is that the amount of protein in a band correlates fairly well with the amount of protein stain absorbed during the staining process. As Dejong (1955) pointed out, however, proteins may differentially absorb a stain so that equally stained bands may not have equal amounts of proteins. Interpretation of densitometric data must therefore proceed with caution. One helpful development is that the equipment currently available for densitometry is more sensitive.

Brewer and Sing (1970) listed the following factors that may affect the results of densitometry:

1. Proteins have different affinity for stains.
2. Some isozymes stain as broad instead of discrete bands and therefore have varying staining intensity in the same band.
3. Some stains have thresholds (i.e. a certain number of molecules of dye must be converted before the dye turns color).
4. Differences in gel thickness will affect the amount of protein available for a staining reaction.
5. A precipitate of the dye may form on the gel surface as a result of the enzyme action that may interfere with evaluation of a zymogram.

Some of these problems can be resolved through improved technique and better equipment.

Chapter 5

DATA COLLECTION AND ANALYSIS

Once a gel is stained, it must be scored without a delay. Scoring is a process of gathering information from zymograms. How this is done will depend on the kind of information that is desired. This chapter presents sample data sheets for gathering information on isozyme patterns and discusses the principles of the genetic analysis of electrophoretic data.

5.1. Data Sheets

It is possible to score (or "read") a stained gel without draining out the staining solution from the staining tray. A light table sometimes helps in scoring an average-quality zymogram. For scoring purposes, one may develop and customize an electrophoresis data sheet. When surveying novel germplasm for variants, a squared (quadrilled) paper may be used so that the various patterns can be diagrammed to show the relative locations of all the bands (Figure 5-1). When surveying genetic materials for which previous studies have identified and authenticated isozyme patterns, codes for the various patterns may be used in scoring the new entries [e.g. A and B for two different patterns, F (fast allele), and S (slow allele), etc.] (Figure 5-2).

It may be desirable to record the exact locations of various isozymes on a gel. This is accomplished by determining the distance of migration of a marker dye and using that as a reference point; all other band locations are then determined relative to it. The values obtained are called relative mobility (Rf) values and are only to be considered approximate values, because different electrophoretic conditions can produce a different set of values.

REMINDER

1. Chemicals for preparing gel fixatives produce fumes and must be handled in a fume chamber.
2. Do not discard a gel simply because it is not "perfectly" resolved or stained. For some studies, even poorly stained gels can be successfully scored.
3. Bands vary in size as well as intensity of staining, so in scoring stained gels one should note these differences. More emphasis should be placed on differences in mobility, however, when comparing zymograms to detect different alleles. Noting differences in intensity is especially helpful when investigating heterozygotes of multimeric enzymes and polyploid species.

5.2. Principles of Isozyme Data Interpretation

Certain basic guidelines are generally followed in the interpretation of isozyme data. The type of information extracted from isozyme data will depend on the objective(s) of the researcher. After staining and incubation, a gel displays a visible pattern that is characteristic (fingerprint) of a specific enzyme (that for which it was stained), plant tissue or part, age (maturity), and developmental stage. Distinctly different patterns on a zymogram are described varioulsy as zymotypes, electromorphs, electrophoretic variants, etc. These patterns are phenotypes of their respective enzyme loci.

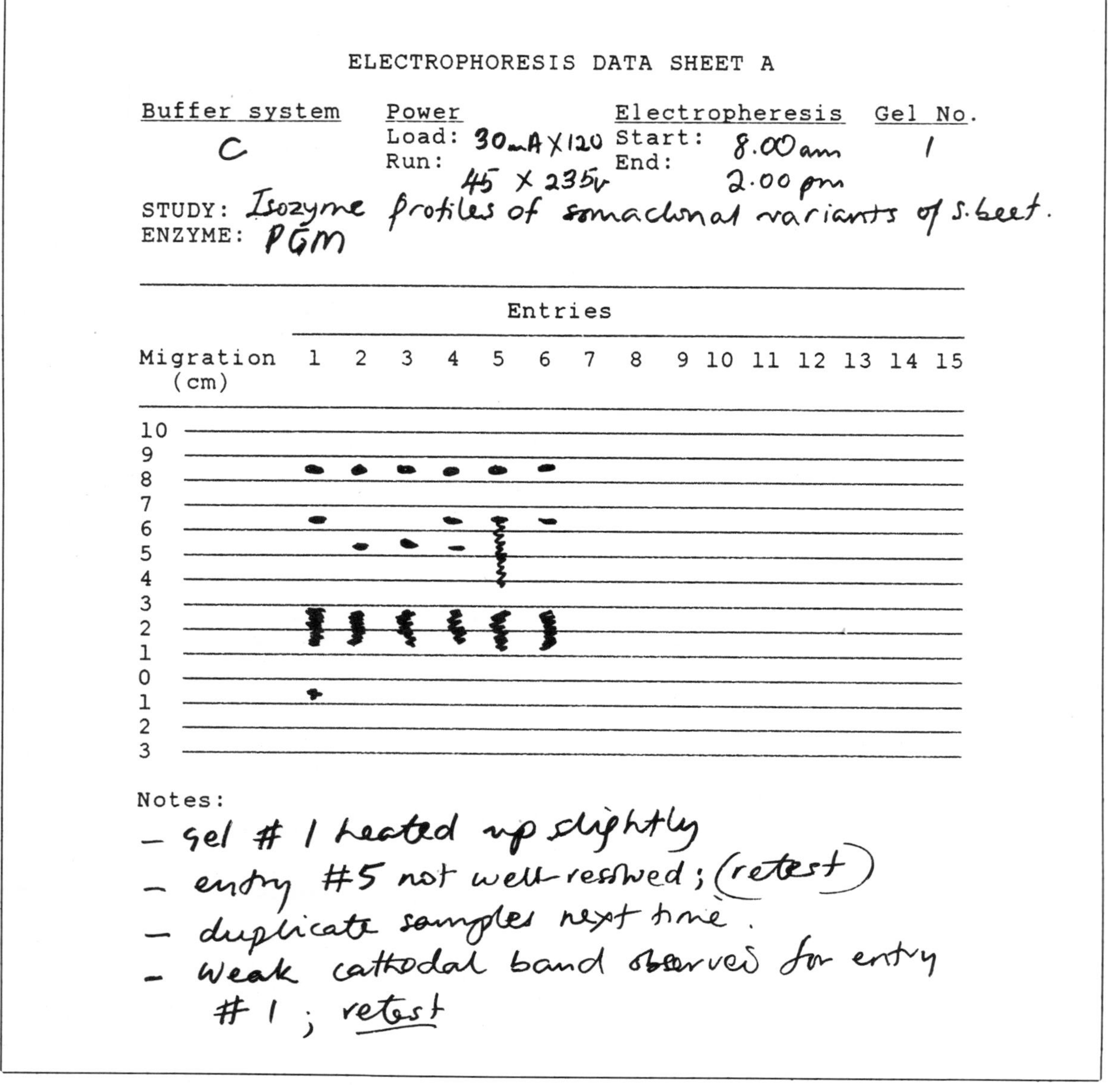

ELECTROPHORESIS DATA SHEET A

Buffer system: C
Power: Load: 30mA X 120; Run: 45 X 235v
Electropheresis: Start: 8.00 am; End: 2.00 pm
Gel No.: 1

STUDY: Isozyme profiles of somaclonal variants of S. beet.
ENZYME: PGM

Entries

Migration (cm) 1 2 3 4 5 6 7 8 9 10 11 12 13 14 15

10
9
8
7
6
5
4
3
2
1
0
1
2
3

Notes:
- gel # 1 heated up slightly
- entry #5 not well-resolved; (retest)
- duplicate samples next time.
- Weak cathodal band observed for entry #1; retest

Figure 5-1. An electrophoresis data sheet for recording isozyme phenotypes when investigating an unfamiliar genetic material and/or enzyme. In such a case, drawing the patterns observed from the zymogram may prove useful.

There is some debate as to whether electrophoretic variation should be described in certain genetic terms. For example, King and Ohta (1975) argue that electrophoretically detected variants that are usually described as alleles (i.e. alternative forms of a gene at a *distinct* locus) are actually sequentially heterogeneous allele classes. Despite the similarity in mobility and catalytic effect, these mobility classes may consist of a number of proteins of different primary structure. The particular protein separation technique employed may not effectively distinguish among the component alleles in a class. Therefore, they argue, describing such heterogeneous classes as "alleles," or their frequencies as "allele frequencies," is erroneous. They should instead be described as "electromorphs," since zymogram patterns are phenotypes (King and Ohta 1975). Although King and Ohta's argument is technically correct, Allendorf (1977) submits that substituting the term "electromorph" for "allele" only compounds the very misunderstanding that King and Ohta seek to clarify. The identities of most alleles are determined on the basis of phenotypic effects, including numerous electrophoretic variants that have been shown to reflect allelic variation at distinct loci (Allendorf 1977).

This debate does not diminish the utility of isozyme analysis and should not detract from its application. It serves to remind researchers to be more cautious in their interpretation of electrophoretic results.

Gorman and Kiang (1978) classified the patterns produced by isozyme polymorphisms into (*a*) those

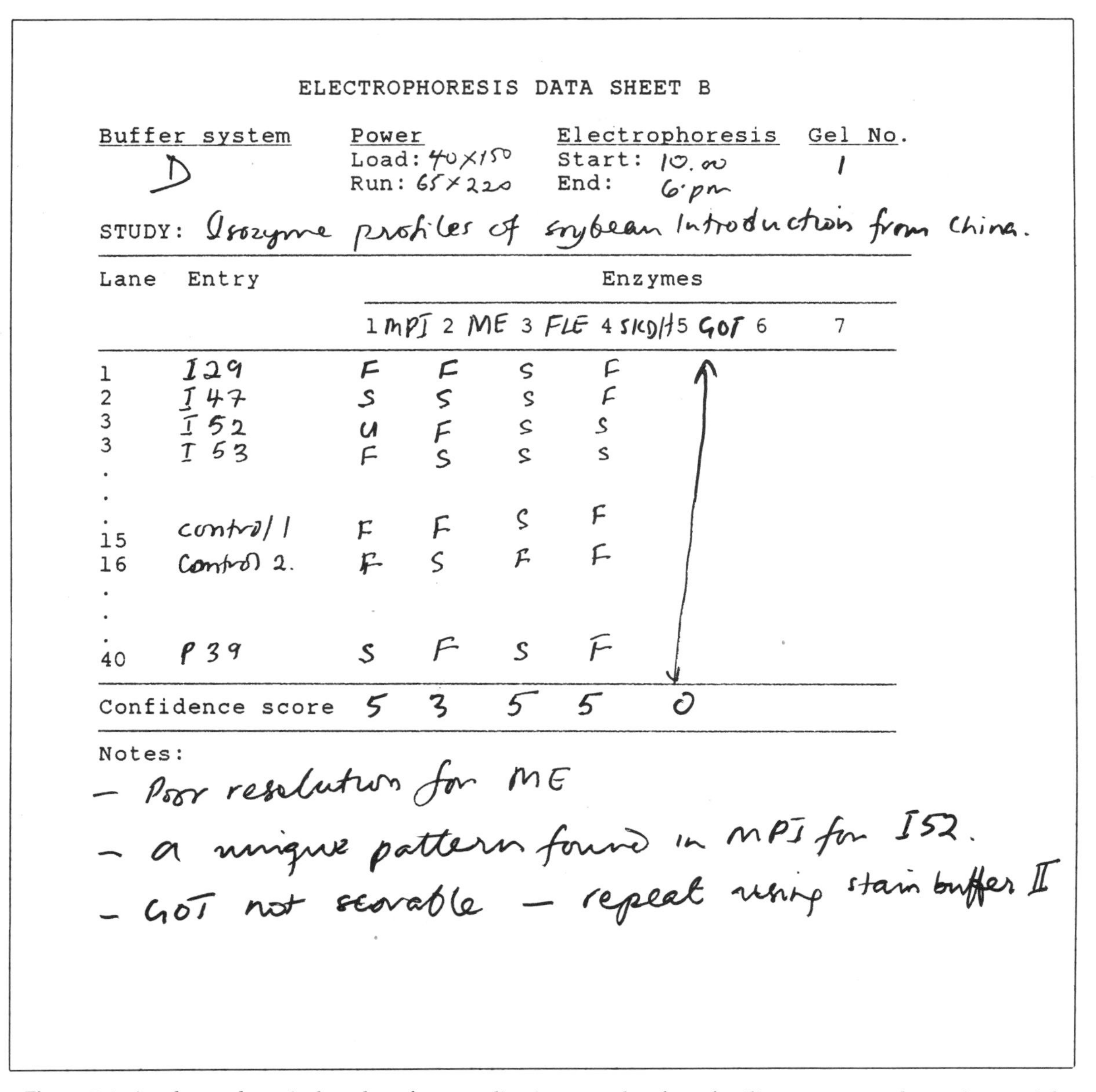

ELECTROPHORESIS DATA SHEET B

Buffer system: D
Power: Load: 40×150 Run: 65×220
Electrophoresis: Start: 10.00 End: 6 pm
Gel No.: 1

STUDY: Isozyme profiles of soybean introductions from China.

Lane	Entry	1 MPI	2 ME	3 FLE	4 SKD/H	5 GOT	6	7
1	I29	F	F	S	F			
2	I47	S	S	S	F			
3	I52	U	F	S	S			
3	I53	F	S	S	S			
.								
.								
.								
15	control 1	F	F	S	F			
16	Control 2.	F	S	F	F			
.								
.								
.								
40	P39	S	F	S	F			
Confidence score		5	3	5	5	0		

Notes:
- Poor resolution for ME
- a unique pattern found in MPI for I52.
- GOT not scorable – repeat using stain buffer II

Figure 5-2. An electrophoresis data sheet for recording isozyme data from familiar enzymes and genetic material. Isozyme phenotypes may be described by codes (e.g. F = fast allele, S = slow allele, and U = a unique, previously unobserved phenotype). A confidence score (scale of 1–5, 1 being least confident) indicates the reliability of the score on the basis of resolution.

with altered mobility in one or more isozymes and (*b*) those with one or more missing isozymes, i.e. null variants. Null variants are often controlled by a single gene; in the heterozygous state, the missing band appears, mimicking dominance gene expression (Pierce and Brewbaker 1978). A null variant arises as a result of changes in the polypeptide that affect enzyme activity (Gorman and Kiang 1978). As mentioned above, the researcher should always repeat an assay before declaring an allele to be null, because a lack of staining does not necessarily mean absence of an active enzyme. Experimental procedures may preclude the expression of an otherwise active enzyme. Sometimes the concentration of an enzyme may be so low that products of its reaction will not be detectable by the protocols used.

Knowledge of the genetic basis of isozyme patterns facilitates their interpretation. It is necessary to authenticate isozymes, for the following reasons:

1. One band is not necessarily the product of a unique isozyme. Synonymous triplets code for the same amino acid (Ayala and Kiger 1980).
2. Similarly, an isozyme may be a molecular hybrid formed by products from two or more enzyme loci. An isozyme therefore is not necessarily a product of a unique locus (Gorman and Kiang 1978). If the active enzyme of an allozyme con-

sists of more than one subunit (one polypeptide chain), the two allelic products are capable of forming a polymeric series of active isozymes (Schwartz and Laughner 1969; Scandalios 1969; Gorman and Kiang 1978; Weeden 1983a).

3. Proteins may sometimes (but infrequently) interact with electrophoretic buffers to produce "fake" bands (Parker and Bearn 1963). Stain artifact bands may also arise from molecular instability (Akroyd 1965). The age of a sample and improper handling prior to electrophoresis may cause some enzymes to change their polymeric state by breaking down into simpler form (e.g. from dimer to monomer) and thereby produce multiple bands (Walter et al. 1965). Bonds between proteins and other charged factors can produce artifact bands (Herbert et al. 1972).
4. Changes in electrophoretic procedure (buffers, pH, overstaining, etc.) may reveal novel bands.

The simplest genetic pattern is produced by an enzyme with one polypeptide chain (*monomer*) in a homozygous individual and two chains in a heterozygous one. One variant has a faster mobility; the other has a slower one. Customarily, the more rapidly migrating band is called the fast (F) allele or band whereas the slowly migrating band is called the slow (S) allele. In a heterozygote, two bands (F and S) will be produced by a monomeric enzyme (Figure 5-3). Other enzymes have more complex quaternary structures comprised of two or more polypeptide chains. An enzyme is a *dimer* when it is composed of two polypeptide chains, because each comprises a different subunit of the enzyme. A homozygous locus specifying a dimeric enzyme is designated as either FF or SS; each band is a *homodimer* (identical subunits). A heterozygous locus will produce three bands: two are homodimers, and the third is a *heterodimer* (unlike subunits) FS, produced by the unlike subunits from the same locus (thus called an *intralocus heterodimer*). This isozyme will often be located midway between the homodimers (Figure 5-4). Bands produced by multiple subunits tend to stain more intensely or may be larger in size (Figure 5-4). DeJong (1955) observed that uptake of the protein stain amido black varied for different proteins, however, so stain intensity is not a conclusive quantitative criterion for distinguishing between isozymes.

Some enzymes are tetrameric, that is they have four polypeptide chains (Figure 5-5). It is important to recognize that heterozygous phenotypes may not always be observed as theoretically expected, due to incompatibility in structural symmetry of the different subunits. An observation of, for example, three bands is not conclusive evidence for a dimeric enzyme but only that the enzyme is polymeric and has at least two polypetide chains. A blurred zone of activity on a zymogram is frequently indicative of a polymeric enzyme. The isozyme patterns for a particular enzyme are not identical from one species to another (Figure 5-6). Furthermore, different quaternary states of the same enzyme may be found in different species (Dixon and Webb 1979) and even in the same species (Gorman and Kiang 1978). A method for predicting the number of bands or the quaternary structure of an enzyme as well as the allelic state of a locus is presented below (section 5.5).

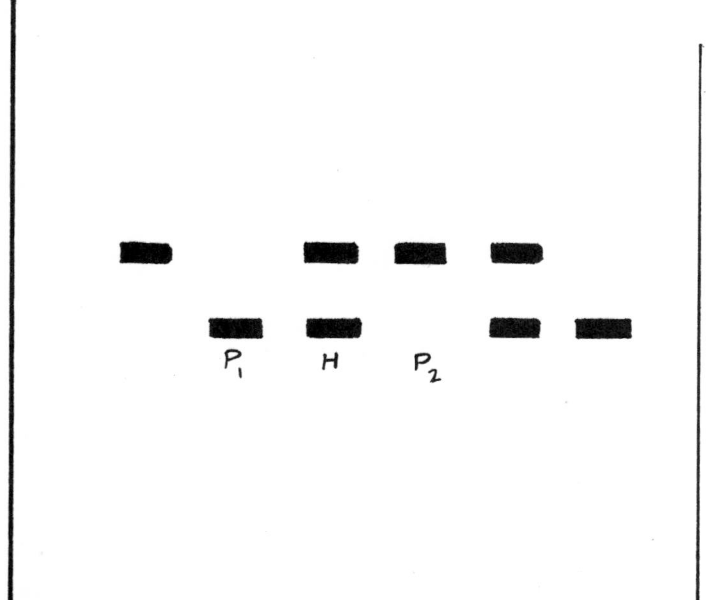

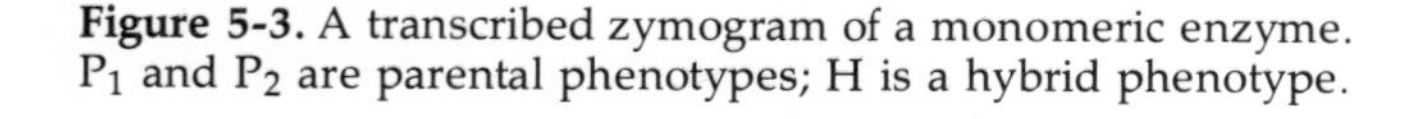

Figure 5-3. A transcribed zymogram of a monomeric enzyme. P_1 and P_2 are parental phenotypes; H is a hybrid phenotype.

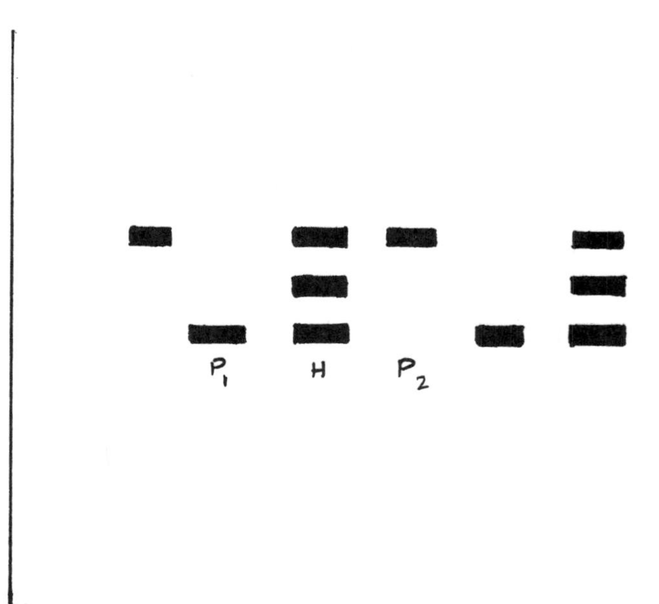

Figure 5-4. A transcribed zymogram of a dimeric enzymes. P_1 and P_2 are parental phenotypes; H is a hybrid phenotype.

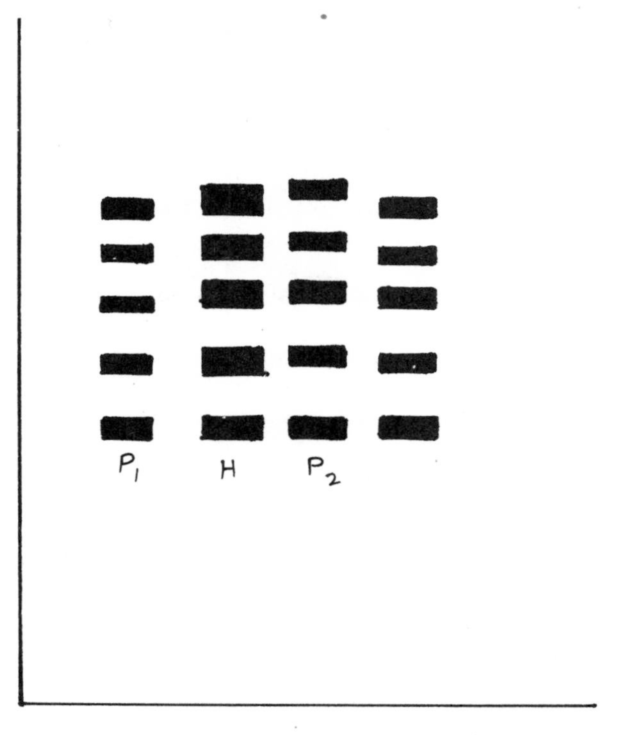

Figure 5-5. A transcribed zymogram of a tetrameric enzyme. P_1 and P_2 are parental phenotypes; H is a hybrid phenotype.

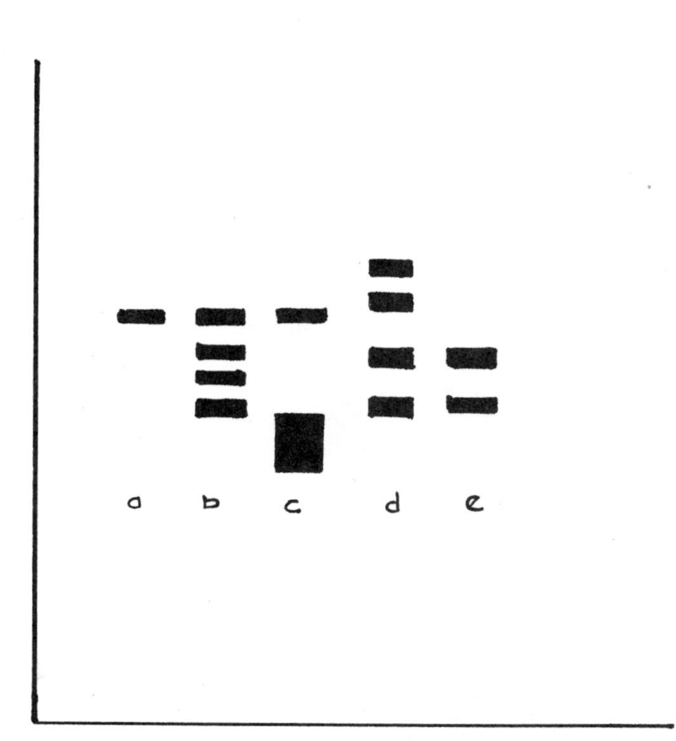

Figure 5-6. A transcribed zymogram showing electrophoretic phenotypes for the same enzyme (GOT) as found in different species and tissues: (*a*) dry bean leaf; (*b*) dry bean seed; (*c*) soybean seed; (*d*) wheat seed; (*e*) sugarbeet leaf.

Gorman and Kiang (1978) listed three guidelines for the successful interpretation of electrophoretic data:

1. Verify the genetic identity of the bands.
2. Consider the molecular substructure of the enzyme.
3. Limit comparisons to samples obtained from the same tissue at the same developmental stage.

Gottlieb and Weeden (1979) described a technique that employs the isozyme pattern from a pollen source to identify the isozymes from somatic tissue. A pollen grain from a diploid organism possesses a haploid complement of chromosomes and will only contain one allele at each locus. Pollen grains from a heterozygous plant will segregate into two classes, each of which will produce a unique subunit of an enzyme. It is not possible for products coded by two alleles of a locus to interact to form an intralocus heterodimer in pollen grains. When pollen extracts are electrophoresed along with extracts from somatic tissues, the pollen will produce a pattern with one missing band (for one heterozygous locus). The isozyme from a somatic tissue located in a position corresponding to the missing isozyme in the pollen grain will be an intralocus hybrid isozyme (Figure 5-7).

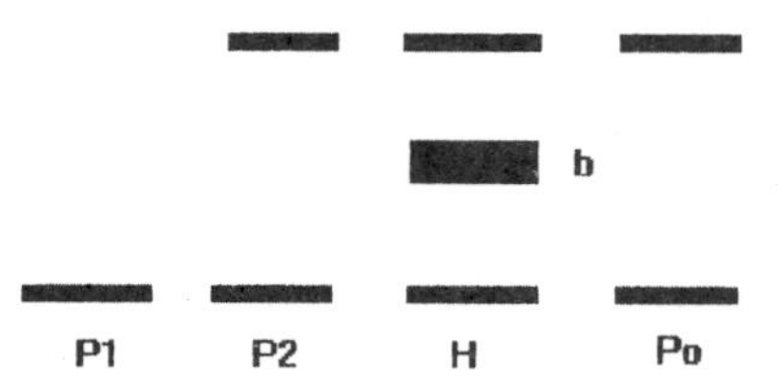

Figure 5-7. Isozyme patterns of parental genotypes, their hybrid and pollen grain. P_1 and P_2 are two parent genotypes, and H and P_o are the hybrids from the two parents and the pollen grain from the hybrid, respectively; b is the intralocus heteromer.

The middle band of the heterozygote (H) represents an intralocus heteromer. Catalytically active enzymes from epistatic sources (interlocus heterodimers) may be obtained in both haploid and diploid tissues, however. Hybrid enzymes will not be formed *in vivo* between subunits of polymeric enzymes located in different compartments of the cell (Gottlieb 1982).

Weeden and Gottlieb (1980) also described a procedure for releasing only cytosolic isozymes by soaking pollen grains in the extraction buffer instead of homogenizing them. By comparing electrophoretic patterns from pollen grains treated as above with those from homogenized somatic tissue, determining the subcellular location of isozymes is possible without employing the conventional fractionation method. Since soaking in a buffer solution generally releases only the isozymes of cytosolic origin, the enzymes located in positions on the gel corresponding to the missing isozymes in the pollen grain will be those from noncytosolic sources (e.g. chloroplast, plastid, and mitochondria).

Exercise caution when interpreting dehydrogenase bands. Racker (1955) observed bands in the absence of a substrate, the so-called "nothing dehydrogenase." This event was attributed to SH groups on proteins that produce zones with strong reducing capacity toward NAD^+ and $NADP^+$. Likewise, be cautious in interpreting isozyme patterns from plants with pathogen infection. Healthy and disease-free tissue must be sampled for enzyme assay. Staples and Stahmann (1963), in examining bean tissue infected with rust, found that one of the isozymes detected derived form the mycellium of the causal fungus. Enzymes with broad substrate specificity (e.g. esterases) are difficult to resolve and also are frequently assayed with artificial substrates not found in natural populations. Such enzymes are best considered as a complex isozymic system (Markert 1977) and interpreted with caution.

5.3. Isozyme Nomenclature

A variety of formats exist for designating alleles at a single locus. Khaler and Allard (1970) designated each allele by its absolute mobility in a particular study. In this system, alleles may be identified as e.g. 5, 5.9, 12.7, etc. Alternatively, alleles may be identified by their mobilities relative to a marker. Alleles also may be designated as F (fast) or S (slow), but more for convenience than accuracy, since this is not suitable when a locus has more than two alleles, unless such categories as "intermediate" (I) and "unique" (U) are introduced. May et al. (1979) used a scoring method that permits the comparison of different gels (i.e. different electrophoretic results) irrespective of differences in absolute mobilities. In this method, the most frequently observed allele at a locus is identified and assigned a mobility value of 100 for its homomeric state. This value becomes the standard or reference-mobility relative to which all other isozyme mobilities are calculated. For example, an allele may be designated as 90 or 110, depending on whether it is more cathodal or anodal relative to the standard allele. For this method of scoring to succeed, the standard genotypes must be included in all the tests to be compared. A variation of the May method was used by Koenig and Gepts (1989), whereby the most common allele of the enzyme is taken as the standard so that all other alleles are assigned numeric values corresponding to the actual distance in millimeters from the standard. The standard is given a score of 100 mm; other alleles may be scored as allele 98, 70, 110, etc.

The earlier literature adopted the same system of identifying loci of isozymes as was used for other genetic loci. The newer system (Rennie et al. 1987) designates an isozyme locus by a three-letter abbreviation of the allele name followed by a number. The abbreviated name of the enzyme is italicized. For example, an aconitate hydratase isozyme locus may be designated as *Aco1*-a, -b, and -c, whereby the number *1* denotes locus 1 and the letters designates the alleles. If an enzyme has more than one locus, the loci will be designated consecutively as *Aco2*, *Aco3*, etc. When the locus exhibits a dominance expression, the two alleles will be designated, e.g. in superoxide dismutase (SOD), as *Sod1* and *sod1* for dominant and recessive alleles, respectively. Some researchers designate a locus with a hyphen between the enzyme name and the locus number (e.g. *Sod-1*). One convention is to identify isozyme loci consecutively from the origin so that the most cathodal locus is locus 1. IUPAC-IUB (1976) recommends that loci be numbered consecutively from the fastest to the slowest bands. In whatever system one uses, consistency is essential so that the same system of numbering is applied to both loci and alleles (e.g. the most anodal locus is 1, and the most anodal allele is a).

5.4. Developing Genetic Models for Inheritance of Isozymes

The first step in building a model for inheritance of isozymes is to identify sources of isozyme variability. This calls for very good electrophoretic procedures so that variants are unambiguously classified. It is possible to distinguish between isozymes that are 1 mm apart (Rick and Tanksley 1983). To do this with confidence, the gel front must be straight and the bands distinct and well-resolved. Next, the different mobility variants must be authenticated by means of planned crosses between appropriate parents. Lines that carry the isozyme patterns of the parents, the various mobility variants, and their hybrid forms should be selected and strategically arranged in a gel and electrophoresed for convenient comparison of band locations and numbers.

After staining, identify zones of activity on the zymogram and test the inheritance of variant bands in these zones by first scoring each band (e.g. as a, b, c) and then testing a hypothesis of a probable genetic system. It is convenient to designate the mobility variants in alphabetical order, the fastest being a. The first hypothesis normally tested is that the mobility variants for each parental band are controlled by a single locus whose alleles are codominantly expressed. Hence, a chi-square test of goodness of fit to the Mendelian genetic ratio of 1:2:1 for codominant gene expression is conducted. If multiple pairs of crosses were made, test each cross type independently, after which crosses may be pooled after a heterogeneity test has been perfomed. Describing all the possible loci an enzyme may possess is not necessary in a single study, since one may not have at hand all the appropriate kinds of variability to accomplish such a purpose. As more genotypes are surveyed, new variants with new alleles or new loci may be discovered.

I use aconitase enzyme (ACO) as an example in model-building because the various steps can be shown in simple form. Models for ACO also have been described by Griffin and Palmer (1987) and Doong and Kiang (1987). The example given here is comparable to those previously reported. Pollen grain samples were not investigated, however, so the location (cytosolic or plastid) of isozymes, as described by Gottlieb and Weeden (1979), will not be possible to tell. Furthermore, the patterns in different tissues and developmental stages were not investigated to find out which bands may be related (by means of observing how they appear or disappear with developmental stage) and hence possibly coded by the same locus. Such practices help in developing models. It is also helpful to transcribe (or photocopy) the zymogram for a gel onto paper along with the R_f values (Figure 5-8).

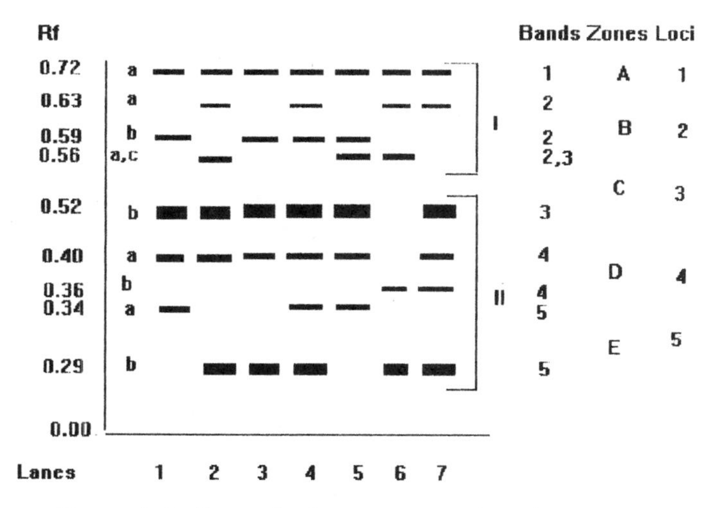

Figure 5-8. Transcribed zymogram of soybean aconitate hydratase isozymes.

GENOTYPES:

Locus	Lanes 1	2	3	4	5	6	7
1	a/a	a/a	a/a	a/a	a/a	a/a	a/a
2	b/b	a/c	b/b	a/b	b/c	a/c	a/a
3	b/b	b/b	b/b	b/b	b/b	a/a	b/b
4	a/a	a/a	a/a	a/a	a/a	b/b	a/b
5	a/a	b/b	b/b	a/b	a/a	b/b	b/b

Two distinct major zones of activity can be identified for ACO, I and II. By comparing the parental banding patterns and band numbers (five) with each other and with the hybrids, a total of five smaller zones of activity (A, B, C, D, E,) can be identified. Each zone contains two variant bands (which may be called a and b), except for zone B, which has three bands (a, b, c), the slowest of which is in the same position as the faster band for zone C (comigration of bands). Zone A has one band with no variants (fixed or invariant locus). The F_2 segregation for pairs of bands in each zone is tested according to the hypothesis of one locus control with codominant alleles. The two mobility variants in band 1 of zone E is used here as an example of how to test the segregation at a locus (Table 5-1). The results show a lack of significant deviation from the 1:2:1 ratio. Zone E may be proposed as a locus, *Aco5*, with two codominant alleles. The other zones and bands are treated similarly. Further, the linkage between different loci is tested for independence. The hybrid pattern shows both parental bands with no intermediate bands. The molecular structure of ACO is likely to be monomeric.

Table 5-1. Genetic analysis of an aconitate hydratase locus of soybean.

Cross	Segregation classes a/a	a/b	b/b	n	χ^2 observed
I	35	75	40	150	.333
II	57	98	45	200	.152
Pooled	92	173	85	350	

Summary

1. ACO5 enzyme is monomeric in the soybean population surveyed.
2. *Aco5* is conditioned by a single locus.
3. *Aco5* locus has two alleles.

The ACO model is not complicated, so the model proposed by Griffin and Palmer (1987) and Doong and Kiang (1987) will probably be acceptable to all researchers. More complicated situations may be encountered that may result in more than one plausible model. The ACO enzyme, according to the genotypes presented in Figure 5-8, shows five features that may be encountered in model building:

1. A nonvariant locus vs. variant locus.
2. Comigration vs. independent migration of bands.
3. Loci with two alleles vs. more than two alleles.
4. Multiple loci.
5. Different band intensities.

The ACO system as presented does not show a null allele. Also, it is possible to encounter skewed segregation in the F_2 due to other genetic problems in the crop.

5.5. Prediction of the Number of Isozymes of Polymeric Enzymes

Polymeric or oligomeric enzymes have multiple structural genes that code for variant subunits of a holoenzyme (an enzyme with the complete complement of subunits). These subunits combine to produce various isozymes. The number of isozymes that can be produced for a diploid organism whose subunits combine randomly with no restriction by subcellular compartmentalization may be predicted from the following mathematical equation (Shaw 1964):

$$i = (s + p-1)!/[p!\ (s-1)!]$$

where i is the number of isozymes (spots or bands on a zymogram), p is the number of polymers (which constitute the polymeric holoenzyme), s is the number of different subunits or variant polymers (in the gene pool of the individual that can interact to form the holoenzyme). [The reader may consult Brewer and Sing (1970) and Dixon and Webb (1979) for further discussion of the subject.] As mentioned above, isozymes may originate from the same locus or different loci. The i value represents the number of nonidentical alleles of a particular enzyme an individual possesses. It follows from the formula that irrespective of the quaternary structure of an enzyme, only one spot or band will be produced in the homomeric (only one kind of interacting subunit) state. Where two different subunits occur, e.g. a heterozygote at a single locus, a monomeric enzyme will display 2 bands, a dimer 3, and a tetramer 5. As the number of different interacting subunits increases (e.g. several loci controlling isozyme phenotype coupled with heterozygosity), the number of bands in a phenotype increases. For example, if two loci are involved and the enzyme is a dimer, there will be four interacting subunits, which upon random association will produce 10 bands. A similar scenario with a tetrameric enzyme will produce 35 bands!

Based upon the assumptions (random combination of polypeptide units, no restriction by subcellular compartmentalization) made by Shaw (1964), one can expect the frequency of different allozymes to be governed by Mendelian laws. If an enzyme consists of two polypeptide chains (dimer), for example, a cross between two parents homozygous for either allele of the enzyme (P_1 = aa, P_2 = a'a') should produce a heterozygote with three isozymes or bands whose genotypes will be aa, aa', and a'a'. Furthermore, the isozymes in the population will have a frequency of 1:2:1, respectively. All things being equal (subunits produced in same amounts and of uniform activities, coupled with good electrophoretic conditions), the frequency of the isozymes will be reflected in the band characteristics (size, location, and intensity). Giving p and s values, it is possible to predict the relative intensity of bands in a phenotype by simply expanding the multinomial. The general formula for a multinomial is:

$$(N!/k_1!k_2! \ldots\ldots k_n!)P_1^{k1}P_2^{k2} \ldots\ldots P_n^{kn}$$

where $N = k_1 + k_2 + \ldots\ldots + k_n$, P_n is the gene dose (or number of copies of an allele), and k_n is the number of the protein products of alleles that produce each band. A value may be calculated for each band in an isozyme phenotype by using this formula.

Figures 5–9 to 5–16 show the types of zymograms expected theoretically for some of the more commonly encountered combinations of genetic control, types of alleles at controlling loci, and quaternary structures of enzymes. The diagrams have not been drawn to scale but show approximate patterns. Many bands often appear as roundish or oval spots. The rectangular shapes in these examples are used just for convenience. I must emphasize again that electrophoretic conditions, the nature of the enzyme, and the physiological state of the organism, among other factors, may cause certain bands to be absent or stained to a lesser intensity.

The following examples involve two-locus control of an enzyme (i.e. the polypeptides or subunits available for random combination are contributed by two loci). Furthermore, the symmetry of a zymotype pattern for a heterozygote will depend upon the allelic state of the parents with respect to the enzyme locus (i.e. the parents could be homozygote + homozygote, homozygote + heterozygote, or heterozygote + heterozygote). The researcher may therefore observe one of three typical patterns for heterozygotes, one of which will be identical to the corresponding pattern for a heterozygote under single-locus control (the two parents make equal contribution to the heterozygote). When the two loci share common alleles, the heterozygote will be expected to have the symmetry of the single locus.

The above exercise may be repeated for different combinations of quaternary structure and numbers of loci and alleles. Formulating expected patterns will help in developing genetic models for inheritance of an enzyme. It must be stressed that these are only expected patterns, however, since they may not always be observed exactly as expected with respect to intensity and number of bands. Each locus may be expressed at different rates or even have different electrophoretic environmental requirements, especially when multiple loci are involved, thus causing atypical patterns. For example, the five bands expected for a tetramer in sugar beet, as was proposed for malic enzyme by Aicher (1988), are not always observed. Frequently, the two lower (more cathodal bands) are not observed.

The examples given so far are relatively simple. As the number of alleles and loci increase, developing models for electrophoretic patterns becomes more complex. Two examples show the increased complexity of model building when there are three alleles not shared in common by the multiple loci. Despite the increase in complexity, the principles followed for the simpler examples do not change.

Aicher (1988) postulated two-locus control and a tetrameric quaternary structure for malic enzyme in sugar beet. The two loci share two of three alleles in common (i.e. there are three distinct subunits as opposed to two in the simpler examples). The above

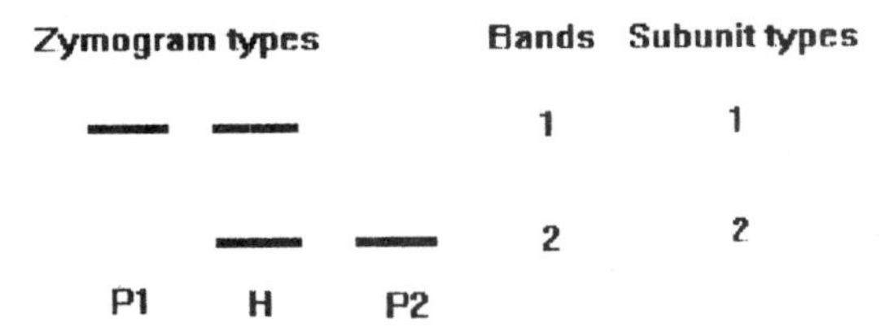

Figure 5-9. Expected zymogram types for an enzyme whose quaternary structure is monomeric and is controlled by one locus with two alleles. P_n and H are parents and their heterozygote, respectively.
GENOTYPES: aa, aa', a'a'. Ratio of band intensities in corresponding phenotypes: 1, 1:1, 1. Number of different interacting subunits: 2. Expected number of bands in the heterozygote: 2. P_n and H are homozygote parents and their heterozygote, respectively.

Zymogram types
Bands Subunit types
1 1,1
3 1,2
3 1,3
P1 H P2

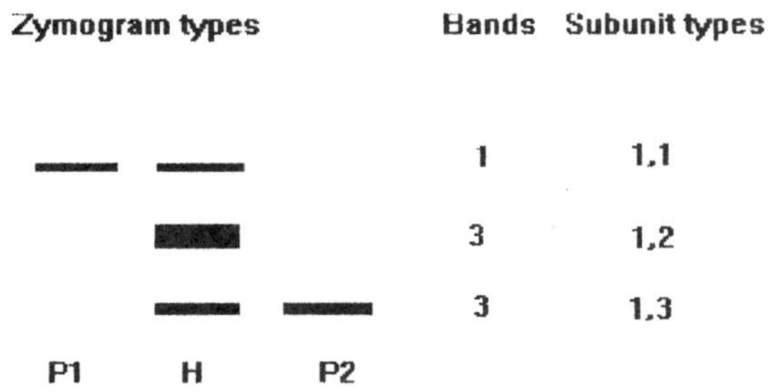

Figure 5-10. Expected zymogram types for an enzyme whose quaternary structure is dimeric and is controlled by one locus with two alleles. P_n and H are the parents and their heterozygote, respectively.
GENOTYPES: aa, aa', a'a'. Ratio of band intensities in corresponding phenotypes: 1, 1:2:1, 1. Number of different interacting subunits: 2. Expected number of bands in the heterozygote: 3. P_n and H are homozygote parents and their heterozygote, respectively.

Zymogram types
Bands Subunit types
1 1,1,1,1
2 1,1,1,2
3 1,1,2,2
4 1,2,2,2
5 2,2,2,2
P1 H P2

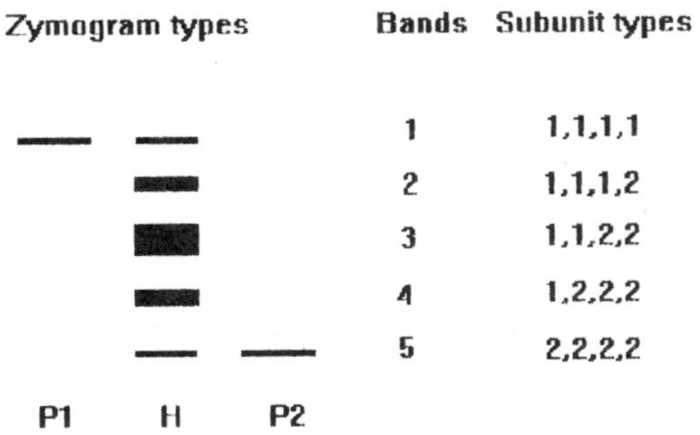

Figure 5-11. Expected zymogram types for an enzyme whose quaternary structure is tetrameric and is controlled by one locus with two alleles. P_n and H are the parents and their respective heterozygotes.
GENOTYPES: aa, aa', a'a'. Ratio of band intensities in corresponding phenotypes: 1, 1:4:6:4:1, 1. Number of different interacting subunits: 2. Expected number of bands in the heterozygote: 5. P_n and H are homozygote parents and their heterozygote, respectively.

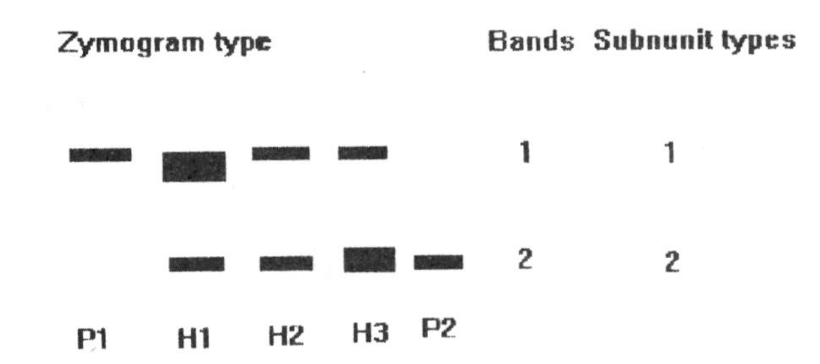

Figure 5-12. Expected zymogram types for an enzyme whose quaternary structure is monomeric and is controlled by two loci with two alleles in common.
GENOTYPES: aaaa, aaaa', aaa'a', aa'a'a', a'a'a'a'. Ratio of band intensities in corresponding phenotypes: 1, 3:1, 1:1, 1:3, 1. Number of different interacting subunits: 2. Expected number of bands in the heterozygote: 2. P_n and H_n are homozygote parents and their heterozygotes, respectively.

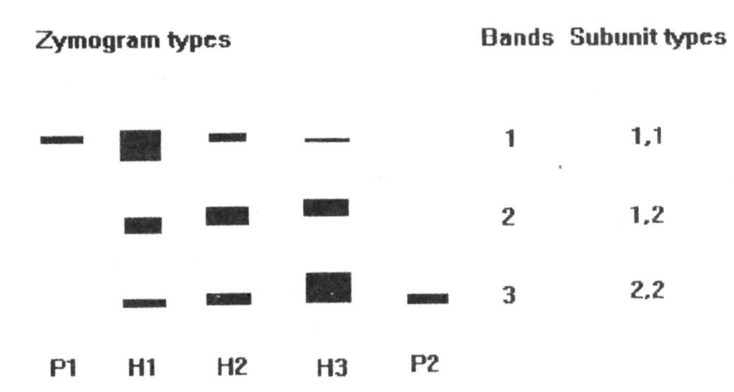

Figure 5-13. Expected zymogram types for an enzyme whose quaternary structure is dimeric and is controlled by two loci that share two common alleles, where P_n and H_n are the parents and heterozygotes, respectively.
GENOTYPES: aaaa, aaaa', aaa'a', aa'a'a', a'a'a'a'. Ratio of band intensities in corresponding phenotypes: 1, 9:6:1, 1:2:1, 1:6:9, 1. Number of different interacting subunits: 2. Expected number of bands in the heterozygote: 3. P_n and H_n are homozygote parents and their heterozygotes, respectively.

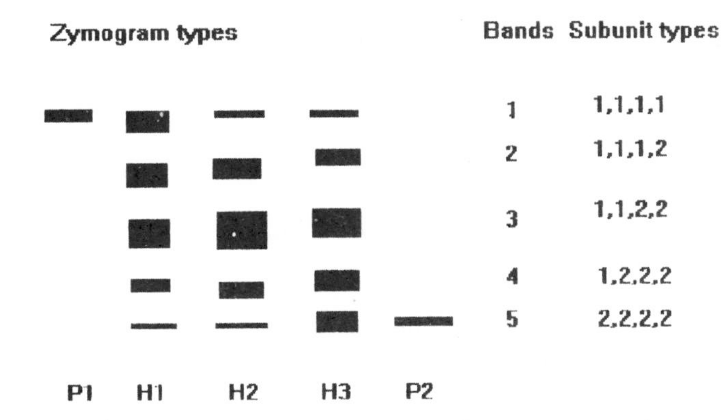

Figure 5-14. Expected zymogram types for an enzyme whose quaternary structure is tetrameric and is controlled by two loci that share two common alleles. Where P_n and H_n are the parents and their heterozygotes, respectively.
GENOTYPES: aaaa, aaaa', aaa'a', aa'a'a', a'a'a'a'. Ratio of band intensities in corresponding phenotypes: 1, 4:4:5:2:1, 1:4:6:4:1, 1:2:5:4:4, 1. Number of different interacting subunits: 2. Expected number of bands in the heterozygote: 5. P_n and H_n are homozygote parents and their heterozygotes, respectively.

situation can arise if a mutation at one of the loci produces a heterozygous condition (Brewer and Sing 1970). Figure 5-15 presents the model for the above scenario. Under very good electrophoretic conditions, the individual bands can be seen on the zymogram (G. Acquaah, unpublished data).

Another three-allele situation was described by Gorman (1983) for *Idh* of soybean. For simplicity, only the cytosolic IDH isozymes are used in the model for this example. The proposed model will also fit a dimeric enzyme coded by one locus with three alleles (Figure 5-16). For a tetrameric enzyme coded by one locus with three alleles, one would expect the typical five bands whereas three bands occur in the IDH example. In model building, look for the simplest model to explain the observed zymogram.

A deviation from the theoretical expectation may occur. Often, the condition of random association among units may not be satisfied (particularly when there is compartmentalization), in which case the holoenzyme will be formed from identical subunits (Brewer and Sing 1970). Instead of five bands for a tetrameric enzyme from two homozygous loci, identical subunits from either locus will interact to produce just two homomeric isozymes (aaaa and bbbb). As mentioned above, electrophoretic conditions may not always be conducive to the appearance of all bands, and thus familiar patterns may not always be observed.

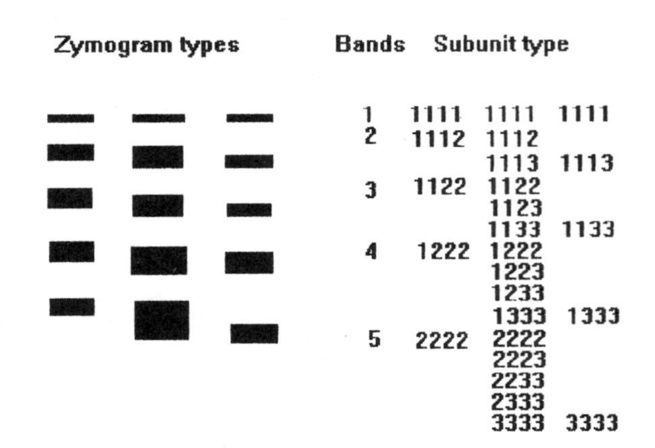

Figure 5-15. Proposed genetic model for an enzyme whose quaternary structure is tetrameric and is controlled by two loci that share one allele in common. P_1 and P_2 are the two parents and H their heterozygote, respectively.
GENOTYPES: aaa'a',aaa,a', aaa,a,. Number of different interacting subunits: 3. Expected number of bands in the heterozygote: 5. P_n and H are homozygote parents and their heterozygote, respectively.

Figure 5-16. Proposed genetic model for an enzyme whose quaternary structure is dimeric and is controlled by two loci with three alleles. The corresponding genotypes: 1=A'A'A'A', 2=A'A'A'A, 3=A'A'AA, 4=A'AAA, 5=AAAA, 6=AAAA'', 7=AAA''A'', 8=AA''A''A'', 9=A''A''A''A'', 10=A'A''A''A'', 11=A,A,AA''A'', 12=A'A'A'A'', 13=A'A'AA'', 14=A'AAA'', 15=A'AA''A''.

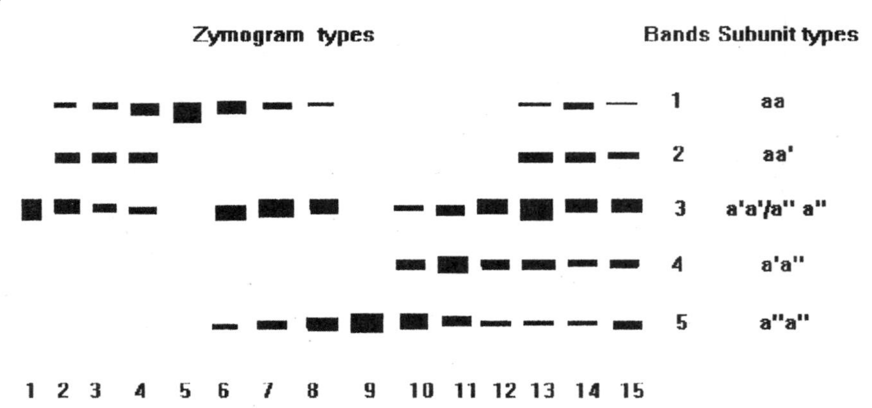

5.6. Trimeric Enzymes

Enzymes with trimeric subunit structure have not been widely reported in either plant or animals. Hopkinson et al. (1976) studied 100 enzyme loci in man and reported 29 monomeric, 43 dimeric, 24 tetrameric, and only 4 trimeric enzyme quaternary structures. When they occur, the expected pattern would be as shown in Figure 5-17 (four bands). When more than two alleles are present in the population, the equivalent pattern of Figure 5-16 would be expected.

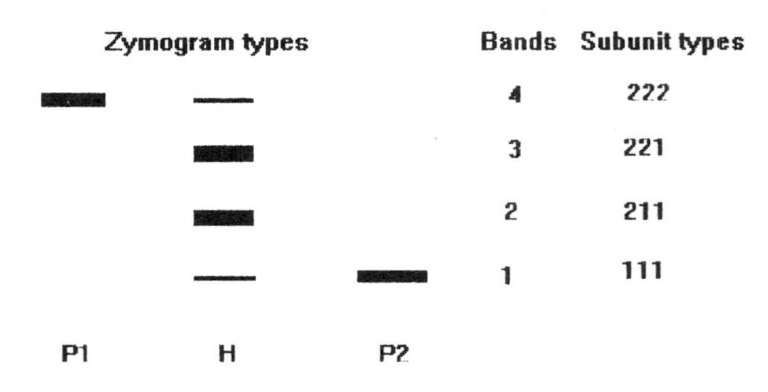

Figure 5-17. Expected zymogram types for an enzyme whose quaternary structure is trimeric and is controlled by one locus with two alleles.

5.7. Isozymes of Tetraploids

The examples given so far have been drawn from studies involving diploid species. A diploid organism, by definition, has two homologous chromosome sets per cell and can have no more than two alleles per locus. This makes the genetics, cytogenetics, and hence isozyme analysis of diploids less complex than those of species of higher ploidy levels. Over 30% of all flowering plants are estimated to be polyploids, that is, their gametic chromosome numbers are multiples of the basic (x) number of their respective genuses. A tetraploid is perhaps the most successful natural polyploid (deWet 1980). The primary purpose of this section is to highlight some of the factors that make the isozyme analysis of polyploids generally more complex than that of diploids.

There are two kinds of polyploids, each with genetic and chromosomal characteristics: *autopolyploids* and *allopolyploids*. Cytogenetically, autopolyploids consist of multiples of the same genome (e.g. AAAA) and exhibit random pairing of homologues during meiosis. Multivalent formation is a common feature of autopolyploidy but bivalents may occur, depending on the species. Genetically, autopolyploids generally exhibit polysomic inheritance (because each chromosome is represented three or more times), multiple allelism (more than two alleles per locus), and gene dosage effects (multiple copies of the same allele at a locus).

Allopolyploids, cytogentically, consist of genomes of divergent species (e.g. AABB) and have chromosomes that are either homologous (chromosomes identical with respect to constituent genetic loci and visible structure) or homeologous (partially homologous) with respect to each other. They exhibit preferential pairing (i.e. bivalents are formed between the most homologous chromosomes). Genetically, allopolyploids exhibit disomic inheritance.

The cytological and genetic characteristics described above are for simple case scenarios. Many gradations occur between the two groups. One rare, noteworthy event is double reduction (chromatid segregation), which may occur during multivalent formation. Double crossover may occur between chromatids at meiosis, resulting in the formation of rare novel genotypes that may complicate analysis. Allopolyploids form the largest group of polyploids and include such commercially important crops as wheat, cotton, and tobacco (Bingham 1980). Important autopolyploids include alfalfa and potato.

Due to the complexity of polyploid genetics and cytogenetics, additional terminology is required to describe various genotypes arising from the combinations of alleles at a given locus. In autotetraploids, special terminologies are often used to describe specific diallelic genotypes on the basis of the number of times a dominant allele is represented: aaaa = nulliplex, Aaaa = simplex, AAaa = duplex, AAAa = triplex, and AAAA = quadruplex. Where multiple alleles occur, the terminologies above are modified (e.g. monoallelic = $a_ia_ia_ia_i$; diallelic = $a_ia_ia_ia_j$; tetra-allelic = $a_ia_ja_ka_l$).

Nielsen (1985) presented all the theoretically possible 35 isozyme patterns of a tetraploid rye grass involving a dimeric enzyme locus (*Pgi–2*) for a one-locus-4-allele case (Figure 5-18). Figure 5-18 shows the effects of multiple alleles and gene dosage on electrophoretic phenotypes. Although some researchers prefer to analyze isozyme data of polyploids by grouping on the basis of electrophoretic phenotype (i.e. as either homozygous or heterozygous), others score and classify entries on a zymogram by identifying specific genotypes on the basis of relative staining intensities. The latter requires that all the bands be clearly visible. As Carr and Johnson (1980) observed, however, concentrations of proteins from unique variants may be very low as compared to those

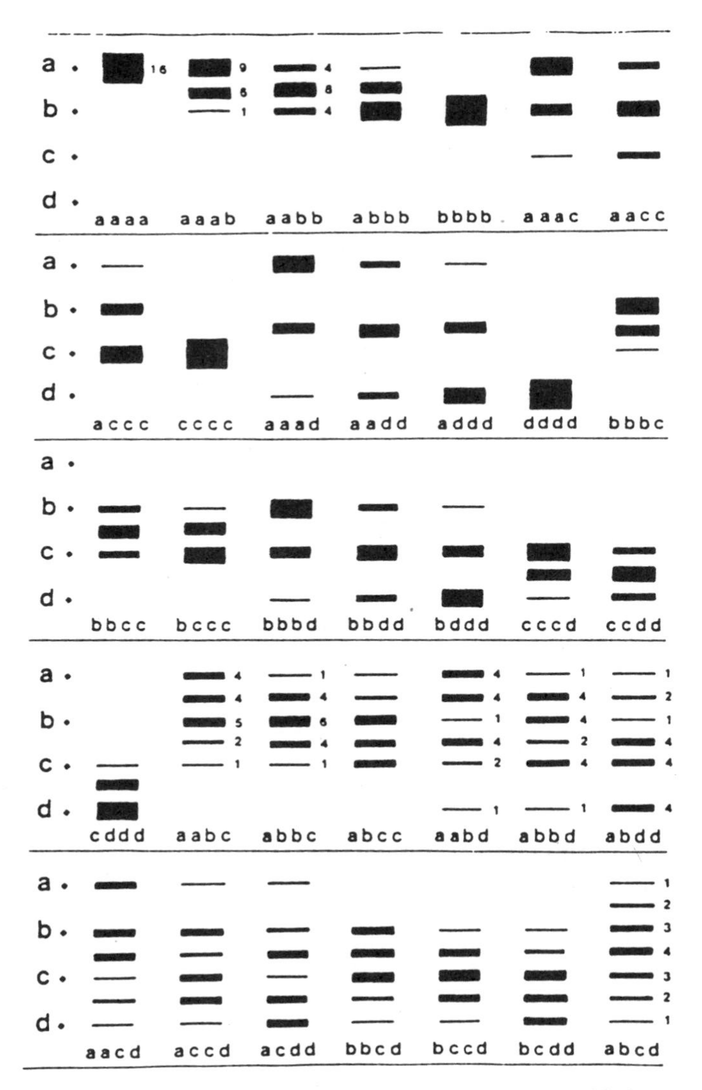

Figure 5-18. Isozyme patterns of tetraploid ryegrass with four alleles in the *Pgi2* locus and the theoretically expected intensity of bands. (From Nielsen 1980 with permission.)

Table 5-2. Expected segregation groups of balanced heterozygotes in diploids, autotetraploids, and allotetraploids. The autotetraploid model assumes no crossovers and random combining of homologous chromosomes.

Parental genotype	Possible gametes	Ratio	Gametic frequency	Zygotic phenotypes in F_2	Zygotic frequency	Heterozygote frequency
Allo-tetraploid						
AAaa	Aa	1	1.00	AAaa	1.00	1.00
Auto-tetraploid						
	AA	1	0.17	AAAA	0.03	
	Aa	4	0.66	AAAa	0.22	
	aa	1	0.17	AAaa	0.50	0.94
				Aaaa	0.22	
				aaaa	0.03	

from the more predominant genotypes. This inbalance will be reflected in the relative intensities of the bands and may lead to an incorrect genotype identification if the electrophoretic conditions are not sufficiently stringent for excellent resolution of all the expected bands. The complexity of isozyme patterns and hence the problems associated with staining intensity as a diagnostic tool will increase as the species ploidy level and quarternary structure of enzymes increase. Carr and Johnson (1980) recommend using enzymes with low subunit number in research involving polyploids whenever possible.

One use for isozyme analysis in the study of polyploids is to determine the mode of polyploidy in an organism. Apart from cytogenetic evidence, auto- and allopolyploids may be distinguished on the basis of inheritance studies. Inheritance studies may not always be conclusive, but cytogenetic data can be tedious to collect and perhaps less conclusive in distinguishing between the two kinds of polyploidy, partly because the pairing mechanisms sometimes produce aberrant ratios in segregational analysis. Isozyme analysis appears to be more effective in distinguishing between the two kinds of polyploids (Tanksley and Rick 1980; Quiros 1982; Krebs and Hancock 1988; Soltis and Soltis 1988; Weeden and Wendel 1989).

Weeden and Wendel (1989) observed that for a single locus, the expected segregational ratios for the two kinds of polyploids are different in all cases except for that of (*a*) a heterozygous locus in an allopolyploid (parental genotype AaBB) and (*b*) chromosomal segregation involving an unbalanced autotetraploid (parental genotype AAAa). Both cases produce a ratio of 1:2:1 in the F_2 zygotic phenotypes. Table 5-2 also shows that one can distinguish between the two kinds of polyploidy on the basis of isozyme segregational data.

5.8. Analysis of Isozyme Population Data

This section very briefly introduces some of the procedures of statistical analysis of isozyme data and provides some references for further detailed consultation. Several good commercial and noncommercial computer softwares are available for these analyses.

Many applications of isozyme analysis are to identify variation and make simple comparisons of different genotypes on that basis. As Weeden (1989) points out, sometimes all that is needed for applying isozyme analysis is that the variation be consistent and environmentally stable. It is not crucial that the genetic basis of isozyme phenotypes be directly ascertained, but they may be assumed to reflect the genetic changes that are observed. Examples of applications based on genetically unauthenticated isozyme variation include cultivar identification and estimation of genetic diversity in germplasm collections (Weeden 1989).

On the other hand, there are many applications that depend on genetically defined variation, such as confirmation of hybridity, marking of monogenic traits, and somaclonal variation, and multilocus applications such as heterosis and heterozygosity, determination of ploidy level, analysis of quantitative trait loci, and development of linkage maps (Weeden 1989). Many of these applications require a knowledge of allelic frequencies. Genotypic frequencies may be obtained from electrophoretic data for various isozyme loci, from which allelic frequencies may then be calculated, based on the codominant expression of isozymic loci.

The allelic frequencies obtained may then be plugged into various equations to obtain estimates of parameters for understanding population differentiation and structure. Several measures of population structure are based on single loci, some of which are introduced in this section. By capitalizing on some properties of isozymes (lack of epistasis or pleiotropy), one can perform several multilocus analyses (Weeden 1989).

5.8.1. Hardy-Weinberg Equilibrium

When isozyme polymorphism is studied in the context of a population of individuals, knowledge of the frequency of allelomorphs and the genotypes that contain them is informative. In a Mendelian population (sharing a common gene pool and freely interbreeding), the genotypic (zygotic) frequencies of the progeny can be predicted from the allelic (gametic) frequencies of the parental gene pool. Furthermore, the gametic and zygotic frequencies remain unchanged from generation to generation (i.e. the population is in genetic equilibrium). The preceding statements will be true only if the following conditions are met, however: (*a*) the population is infinitely large, closed, and in panmixis (randomly mating), and (*b*) there is no selection or differential reproduction, and mutation occurs in one direction. The biometrical model that underlies such population behavior is embodied in the Hardy-Weinberg law.

For a Mendelian population of a diploid organism in Hardy-Weinberg equilibrium, a locus will have two alleles (e.g. A, a), whose frequencies in the population will depend on the genotypic frequencies of the parents that contributed gametes to the gene pool. If the percentages of the two alleles, A, a, in the gene pool are designated as p and q, respectively ($p + q = 100\%$ or 1), the genotypic frequency in the next generation following random mating is given by $p^2 + 2pq + q^2 = 1$ for the corresponding genotypes AA Aa aa.

5.8.2. Allele Frequency

A convenient quantitative measure of genetic variation of a population is the *allele frequency*, which is defined as the proportion of all the alleles at a locus that are of one form or the other of a contrasting pair. Similar to other estimates of population parameters, a large sample size (more than 100) is required to obtain an estimate close to the true population frequency. This measure of genetic diversity is described as a measure of evenness (Brown and Weir 1983), which means that a sample with one frequently occurring allele and three occasionally observed (rare) alleles (a total of four) is less diverse than a sample with four alleles that occur with about the same frequency.

5.8.3. Average Number of Alleles per Locus

The total number of alleles observed at all the enzyme loci investigated, divided by the total number of loci, yields the average number of alleles per locus. Before proceeding with this estimate, it may be best to authenticate each allele (allelism test). This may be achieved with the technique of Gottlieb and Weeden (1979) described above to identify the number of alleles. When interlocus interaction is present, however, this method will run into problems. An estimate of the number of alleles per locus is a measure of the allelic richness (Brown and Weir 1983). For example, if a sample of equal size is drawn from each of two populations A and B, and the average number of alleles are calculated to be 2.15 and 2.39, respectively, population B will be considered more diverse (more distinct kinds of alleles) than population A.

5.8.4. Gene Frequencies of a Population in Hardy-Weinberg Equilibrium

The examples given in this section represent the simplest types of allele systems but can be extrapolated to more complex situations. Some systems are more commonly encountered in isozyme studies than others.

5.8.4.1. Autosomal Locus in a Two-allele System

1. Codominant alleles. Most isozyme systems have codominant alleles (i.e. homozygotes and heterozygotes are phenotypically distinguishable). If a total of N individuals are investigated and the results show X individuals to be homozygous for one allele (FF), Y homozygous for the other allele (SS), and H heterozygotes (FS), it follows from the Hardy-Weinberg (H-W) law that $N = X + Y + H$. A diploid (two alleles per locus) will have a total of $2N$ alleles in the population for the locus in question. An individual that is homozygous for an allele (e.g. FF) will have two copies of that allele but only one copy if it is heterozygous (FS). If the frequency of the F allele is p and that of the S allele is q, the total number of alleles in the population may be calculated as

   $$p = (2X + H)/2N = (X + \tfrac{1}{2}H)/N$$

   and, similarly,

   $$q = (2Y + H)/2N = (Y + \tfrac{1}{2}H)/N.$$

2. One Dominant Allele and One Recessive Allele. Isozyme systems that have null alleles have one dominant and one recessive allele. Homozygous dominant genotype (FF) is not phenotypically distinguishable from a heterozygous (FS) one (unless by means of a testcross). On the other hand, the genotype of the homozygous recessive (ff) is identifiable and can be used to determine the allelic frequencies in the population, provided H-W equilibrium exists. The frequency of ff is then q^2 so that the frequency of f is $\sqrt{q^2} = q$. Since $p + q = 1$, the frequency (p) of the dominant allele (F) is given by $p = 1 - q$.

5.8.4.2. Autosomal Locus in a Multiple Allelic System

1. Multiple Codominant Alleles. In the simplest example of a multiple allelic system (three alleles), there will be six possible phenotypes produced by six genotypes. The allelic frequency is calculated for each allele as was previously done for the two-allele example. The frequencies of the alleles sum up to unity ($a_1 + a_2 + a_3 = 1$) and the frequency of allele a_1 is obtained as $a_1 = (2A + \tfrac{1}{2}H)/2N$.

2. Two Codominant Alleles and One Recessive (Null) Allele. If a locus B has three alleles ($B_1 = B_2) > b$, the frequencies can be designated as p, q, and r, respectively. From H-W law,

$$(p + q + r)^2 = 1 \text{ and}$$
$$p^2 + 2pr + 2pq + q^2 + 2qr + r^2 = 1,$$

for the genotypes B_1B_1, B_1b, B_1B_2, B_2B_2, B_2b, bb, and the phenotypes B_1 ($B_1B_1 + B_1b$), B_1B_2, B_2 ($B_2B_2 + B_2b$), and b.

5.8.5. Heterozygosity of a Population

Heterozygosity is the probability that two alleles for a particular locus selected randomly from a population are different in form (Ayala and Kiger 1980). Heterozygosity estimates include only enzymes with codominant expression and hence underestimates heterozygosity because some enzymes may have a different mode of inheritance. For self-pollinating crops, the samples taken for an isozyme study may not be from a panmictic population, thus rendering the estimates of heterozygosity less useful. In situations such as above, an expected average heterozygosity should be calculated.

The expected average panmictic heterozygosity of a population may be estimated with the Nei's (1975) measure, which calculates the expected mean panmitic heterozygosity as: $H_e = 1 - \Sigma f_i^2$ where f is the frequency of the ith allele averaged over the number of variable loci in the data set. This average is a very effective index because it takes into account the average number and frequency of alleles as well as the total polymorphism (Ayala and Kiger 1980).

5.8.6. Percentage of Polymorphic Loci in a Population

Generally, a locus whose most common allele occurs with a frequency of less than 0.99 is considered polymorphic. The 0.95 cutoff point also may be used to define polymorphism. If several populations of the same species are being investigated, an average polymorphism may be calculated. The shortcomings of this index of variability are (*a*) it is arbitrary, because one has to define polymorphism with a cutoff point; (*b*) it does not distinguish between degrees of variation, because in estimating average polymorphism, all loci are weighted equally without regard to how polymorphically different they may be (Ayala and Kiger 1980). The percentage of polymorphic loci provides a very crude estimate of the level of genetic diversity because it incorporates the two basic concepts of diversity (allelic richness and evenness) in general ways (Brown and Weir 1983). In addition, the efficiency of this estimate is affected by the sample size and the number and kinds of enzymes the researcher includes in a study.

5.8.7. Linkage Analysis

It is often desirable to establish the linkage relationships among enzyme loci studied (i.e. failure of genes from two independent loci to independently assort). There is evidence that a single locus cannot be studied independently of closely linked loci (Marshall and Allard 1970; Lewontin 1974). Mapping of quantitative traits requires finding the gene markers linked to them. Mapping of the genome of many crops through use of isozymes is in progress (Tanksley et al. 1982; Rick and Tanksley 1983; Edwards et al. 1987). In isozyme studies, linkage between different loci of the same enzyme or between enzyme loci and morphological trait loci may be investigated with Mather's method (1951). The increasing interest in the use of isozyme markers for mapping quantitative trait loci has led to attempts to detect linkage between enzyme loci and quantitative trait loci [see Tanksley et al. (1982) for a description of one such procedure].

5.8.8. Estimation of Genetic Recombination

Allard (1956) developed equations to facilitate the estimation of recombination fractions by the maximum likelihood method. Sokal and Rohlf (1969) proposed the log-likelihood test statistic (G) that can be used to test the goodness of fit of a progeny distribution to a Mendelian ratio (test of free recombination). Several types of computer software are available for estimating recombination and other segregation analyses, thus rendering consultation of Allard (1956) unnecessary.

5.8.9. Estimates of Genetic Divergence of Populations

Genetic distance between populations can be assessed by several methods, including Mahalanobis' (1936) D^2 statistic, Roger's (1972) distance, and Nei's (1972) distance.

5.8.9.1. Rogers' Distance Rogers distance (R) is a measure of the biological distance between separated groups. $R = 0$ when the two populations are genetically identical; $R = 1$ when the populations are fixed for different alleles (Rogers 1972). One may also calculate R over a number of loci by simply using the arithmetic mean at each locus (Rogers 1972).

5.8.9.2. Mahalanobis' D^2 Statistic The D^2 statistic is often a part of discriminant analysis and is used to indicate the biological distance between groups. As introduced by Mahalanobis (1936), it was used with quantitative morphological traits. The D^2 procedure is similar in principle to the Rogers' (1972) distance but uses allele frequency data.

5.8.9.3. Nei's Statistics Nei (1972, 1973) developed one of the most widely used indices of population diversity. His genetic diversity statistics allow partitioning of variation both within and among different populations. With this statistical procedure, one can estimate the total genetic diversity—genetic diversity due to variation among populations and between populations—and the average genetic distance between populations. The basis of this genetic identity estimate is the probability that two alleles drawn from the subject populations under comparison are the same. The probability of realizing this event is a function of the allele frequency in the subject populations. If two populations have no alleles in common, their genetic identity is 0.0; however, if they have the same alleles in the same frequency, their identity value is 1.0.

5.8.9.4. Wright's Statistic Wright (1951) developed a statistic (F) to measure the deviation from Hardy-Weinberg expectation (i.e. deviation from panmitic genotype distribution). An F value of near -1 indicates an excess of heterozygotes, whereas a value near 0 indicates that the population is in Hardy-Weinberg equilibrium. An estimate of F near 1 is indicative of fixation (highly inbred).

5.8.9.5. Other Statistical Procedures Some of the earlier reports on isozyme studies were entirely qualitative and descriptive. Now studies can be quantitatively analyzed by the use of formal statistical procedures such as analysis of variance. When the assumptions of parametric statistics are seriously violated, nonparametric statistics such as Mann-Whitney U and Wilcoxon's matched pairs test may be used.

Chapter 6

WORKED EXAMPLES

This chapter presents examples of isozyme data manipulation involving some of the statistical methods introduced in Chapter 5. Most of the examples are taken from my own experimental data. I have included several formats for the presentation of results. Computer software is available for many of these analyses.

6.1. Allele Frequency

Fifty clones of sugar beet (*Beta vulgaris* L.) were surveyed for isozyme variability. The example presented here uses the results for four enzymes: malate dehydrogenase (MDH), phosphoglucomutase (PGM), shikimic dehydrogenase (SKDH), and phosphoglucose isomerase (PGI). One form for the analysis is shown in Table 6-1, where F = fast allele and S = slow allele, respectively. Allele frequency is F = 58/100 = 0.58; S = 42/100 = 0.42. Alternatively, the allele frequencies may be calculated from the formula given in Chapter 5 as follows:

Frequency of F $= (X + \frac{1}{2}H)/(N) = (20 + 3)/50 = 0.58$.

Similarly,

Locus *Pgm1*: Allele frequency of F = 61/100 = 0.61, and S = 39/100 = 0.39;
Locus *Pgi1*: Allele frequency of F = 64/100 = 0.64, and S = 36/100 = 0.36;
Locus *Skdh*: Allele frequency of F = 2/100 = 0.02, and S = 98/100 = 0.98.

The above steps are followed for all the loci under investigation. A summary may be prepared as shown in Table 6-2. The allele 1 for SKDH appears to be rare in the sugar beet population. Nontheless, at the 0.99 cutoff point, it can still be said that SKDH is polymorphic in the sugar beet population surveyed.

Table 6-1. Analysis of allele frequency at the MDH locus in a clonal population of sugar beet.

	Genotypic class			
	FF	FS	SS	Total
No. individuals	26	6	18	50
Alleles				
F	52	6	0	58
S	0	6	36	42
F + S	52	12	36	100

Table 6-2. A summary of the allele frequencies at four enzyme loci in a clonal population of sugar beet.

	Allele	
Enzyme locus	1	2
Mdh1	0.58	0.42
Pgm1	0.61	0.39
Pgi1	0.64	0.36
Skdh1	0.02	0.98

6.2. Polymorphism

The polymorphism at each locus is obtained separately. When several populations of the same species have been studied, the polymorphism at all the loci investigated may be summed over all the populations to obtain an estimate of the average polymorphism. In soybean [(*Glycine max* (L) Merr.)], two populations were surveyed for isozyme variability. Seventeen enzyme loci were investigated. The results are presented in Table 6-3. Remember that polymorphims must be defined (i.e. use the 0.99 or 0.95 criterion).

Table 6-3. Analysis of polymorphism at 10 enzyme loci in two soybean populations.

Population	No. poly. loci	Total no. loci	Avg. poly.	% Poly.
A	10	17	0.588	58.8
B	8	17	0.471	47.1
Average			0.529	

6.3. Heterozygosity (H)

Table 6-4 presents the general procedure for estimating heterozygosity (*H*) for populations in which random mating can be assumed to occur. The average heterozygosity of the population of the sugar beet clones surveyed was 17.2%. In practice, many more loci will be included to make the estimate more valid and useful. The *H* estimate for each locus may also be an average if more than one population is investigated.

Table 6-4. Heterozygosity at five enzyme loci in a sugar beet clonal population.

Enzyme locus	No. heterozygous individuals	Total individuals examined	H
Mdh3	4	50	0.08
Mdh1	6	50	0.12
Pgm1	15	50	0.30
Pgi1	18	50	0.36
Skdh	0	50	0.00

Average = 17.2%.

6.4. Expected Heterozygosity (H_e)

Expected heterozygosity (H_e) is estimated for populations in which random mating is not possible. Table 6-5 presents expected heterozygosity estimates for a soybean population. The formula for estimation is $H_e = 1 - \Sigma f_i^2$. For *Idh1*, for example, $H_e = 1 - (f_1^2 + f_2^2) = 1 - (0.723^2 + 0.276^2) = 40$. The other loci are treated similarly (Table 6-5).

Table 6-5. Expected heterozygosity estimates at five enzyme loci of a soybean population.

Locus	Allele frequency		
	F = (f_1)	S = (f_2)	H_e
Idh1	0.723	0.276	0.401
Idh2	0.533	0.466	0.499
Aco4	0.564	0.435	0.493
Pgm1	0.253	0.746	0.379
Sod1	0.923	0.076	0.142

Average estimate is 0.382

6.5. Segregation at a Single Locus

An F_2 population of a sugar beet cross provided the data used in Table 6-6. The segregation classes for the *Idh1* locus were tested for deviation from the Mendelian ratio of 1:2:1 for codominant inheritance pattern. The observed segregation ratio in the above example did not deviate significantly from the expected ratio of 1:2:1. The alleles at the *Idh1* locus are codominantly expressed.

Table 6-6. Analysis of the segregation at the *Idh1* locus of sugar beet.

Cross	Genotypic classes					
	Idh1-a/ Idh1-a	*Idh1-a/ Idh1-b*	*Idh1-b/ Idh1-b*	*n*	χ^2	*P*
138D × AP-1	26	56	22	104	0.923	0.630

$\frac{Idh1\text{-}a}{Idh1\text{-}a} \times \frac{Idh1\text{-}b}{Idh1\text{-}b}$ = genotype of cross

6.6. Linkage Between Enzyme Loci

An evaluation of an independent assortment of alleles at the *Idh1* and *Pgm1* loci was performed by testing the joint segregation of the loci in a dihybrid cross. The ratio tested was 1:2:1:2:4:2:1:2:1 (i.e. 1:2:1 within 1:2:1). The segregation classes are presented in Table 6-7. For linkage analysis, one should indicate the linkage phase (i.e. either coupling or repulsion) in each parent (by inspecting the progeny distribution). Table 6-8 presents an alternative arrangement of the data. The ratio to be tested in Table 6-8 is 1:2:2:4:1:2:1:2:1.

Table 6-7. Contingency table of the joint segregation of *Idh1* and *Pgm1* loci of a dihybrid cross of sugar beet.

	Idh1-a/ Idh1-a	*Idh1-a/ Idh1-b*	*Idh1-b/ Idh1-b*
Pgm1-a / *Pgm1-a*	5	12	6
Pgm1-a / *Pgm1-b*	14	26	11
Pgm1-b / *Pgm1-b*	6	15	4

Chi squared (df = 4) = 0.071, P = 0.898, r = 0.468 +/− 0.049.

Table 6-8. Joint segregations of various gene pairs in different soybean crosses.

Cross	Pair of loci	Phase	Genotypic classes										*n*	χ^2
			AB/AB	AB/Ab	AB/aB	AB/ab	+ Ab/aB	aB/aB	aB/ab	Ab/aB	Ab/aB	ab/ab		
1	*Idh1-Idh2*	R	.	.	.		.	.	.	.	.	.		
2	*Idh1-Pgm1*	R	.	.	.		.	.	.	.	.	.		
3	*Idh1-Aco4*	C												

Cross 1 = MNS 372 × ONT 42 etc., R = repulsion; C = coupling. Dots represent data points.

Table 6-9. Genotypic segregation at an enzyme locus of sugar beet cross.

Cross	Genotypic classes					
	a/a	a/b	b/b	*n*	χ^2	*P*
138D (a/a) × AP-1 (b/b)	.	.	.	.	.	.
.						
.						

Dots represent data points.

Table 6-10. Joint segregation of a locus with dominant expression and another with codominant expression.

Cross	Pair of loci	Phase	AB/AB + AB/Ab	AB/aB + AB/ab + Ab/aB	aB/aB + aB/ab	Ab/Ab	Ab/ab	ab/ab	*n*	χ^2	*p*
			Genotypic classes								
1	.	.	.	.	.	.	.	.	.	.	.
2	.	.	.	.	.	.	.	.	.	.	.
3	.	.	.	.	.	.	.	.	.	.	.

Dots represent data points.

This particular arrangement was chosen for convenience and to create a desired pattern. To make a table even more simple, one may use the allelic designations instead of the entire genotypic description (Table 6-9). The choice of a format for data presentation is entirely up to the individual researcher, as long as the results are reported clearly and understandably.

To test the linkage between loci with different gene expression, the format needs not change, but the ratio to be tested will change. For example, when testing the joint segregation for a locus that shows codominant expression and another that shows dominance, the ratio to test will be 3:6:3:1:2:1 (i.e. 1:2:1 within 3:1). A sample arrangement is presented in Table 6-10.

6.7. Linkage Between an Enzyme Locus and a Morphological Trait Locus

This example uses the sugar beet data to test the linkage between hypocotyl color and the *Idh1* and *Pgm1* loci. In sugar beet, purple hypocotyl color is dominant to yellow. Table 6-11 shows the joint segregation of the gene pair tested against the expected ratio of 3:6:3:1:2:1.

Table 6-11. Contingency tables for the joint segregation of (*a*) hypocotyl and *Pgm1* and (*b*) hypocotyl and *Idh1* loci, respectively, in sugar beet.

(a)	Hypocotyl color R^-/R^-	Hypocotyl color *rr*/*rr*	(b)	Hypocotyl color R^-/R^-	Hypocotyl color *rr*/*rr*
Pgm1-a/*Pgm1-a*	15	8	*Idh1-a*/*Idh1-a*	22	4
Pgm1-a/*Pgm1-b*	36	15	*Idh1-a*/*Idh1-b*	46	10
Pgm1-b/*Pgm1-b*	21	4	*Idh1-b*/*Idh1-b*	8	14
Chi-square = 2.373			Chi-square = 19.170		
P = 0.305			*P* = 0.00006		
r = 0.414 +/− 0.059			*r* = 0.271 +/− 0.049		

Conclusion: *Idh1* and the hypocotyl color loci of sugar beet are linked.

6.8. Multilocus Analysis

Multilocus analysis enables the researcher to examine the collective effect or pattern of isozymes within and between defined biological groups (e.g. species, gene pools). When loci are examined individually, clear patterns or trends may not always be obvious. Multivariate procedures of data analysis are the appropriate tools for handling data with multiple classifications. Some of the multivariate procedures used in isozyme analyses include the following:

6.8.1. Similarity Index

The isozyme profiles for several public releases of soybean (Table 6-12) provide an example of use of the similarity index (SI) (C. Sneller, personal communication). The total number of enzymes examined is 10. The similarity index for Century vs. Corsoy 79 = 6/10 = 0.60 (i.e. similar at 6 out of 10 loci), vs. Pella = 7/10 = 0.70, vs. Hardin = 5/10 = 0.50, and vs. Sibley = 6/10 = 0.60. Corsoy 79 vs. Pella = 5/10 = 0.50, vs. Hardin = 9/10 = 0.90, and vs. Sibley = 6/10 = 0.60. Pella vs. Hardin = 4/10 = 0.40 and vs. Sibley = 5/10 = 0.50. Hardin vs. Sibley = 5/10 = 0.50. Pella and Hardin are the most dissimilar pair whereas Corsoy and Hardin are the most similar pair. A cross between Pella and Hardin would most likely yield useful progeny for the practice of selection.

Table 6-12. Isozyme profiles of some soybean cultivars.

Cultivar	*Idh1*	*Idh2*	*Aco4*	*Mpi*	*Pgm*	*Dia*	*Per*	*Fle*	*Pgi*	*Pgd*
	Enzyme locus									
'Century'	F	F	F	S	S	B	L	S	S	F
'Corsoy 79'	F	S	F	F	S	B	D	S	F	F
'Pella'	F	F	S	S	S	B	L	S	F	S
'Hardin'	F	S	F	F	S	B	D	F	F	F
'Sibley'	F	F	F	F	F	C	L	S	F	F

F, S = fast, slow alleles; B, C, L, or D denote different patterns obtained for the particular enzyme.

6.8.2. Multivariate Analysis

Isozyme data may be subjected to multivariate analysis such as clustering, principal components analysis, etc. Worked examples are not included here because these procedures usually require the use of computers. Many types of computer software are available for such analyses.

6.9. Other Statistical Procedures

Sometimes the data available do not require the aid of a computer to analyze. The following examples help in performing quick estimates of common statistics.

6.9.1. Chi-square Test

The general form of the equation is

$$\chi^2 = \sum_{j=1}^{n} (O_j - E_j)^2/E_j.$$

where O = observed frequency and E = expected frequency of an event. With the data in Table 6-1 for an example, a 2 × 3 contingency table may be set up for analysis (Table 6-13). The ratio being tested for goodness of fit is 3:6:3:1:2:1 = 16. The expected values for the various cells are obtained a follows: For the first cell, $E = (3/16)\ 99 = 18.56$; for the last cell, $E = (1/16)\ 99 = 6.19$.

$\chi^2 = (15 - 18.56)^2/18.56 + \ldots + (4 - 6.19)^2/6.19.$
$\chi^2 = 2.846.$

The degrees of freedom (df) = $(k - 1)(n - 1) = (3 - 1)(2 - 1) = 3$, where k and n = number of columns and rows, respectively.

When goodness of fit of data to a known pattern is not being tested but the association between two classes is desired, the expected values are estimated by using the border totals of the contingency table (Table 6-14). The expected value for cell 1 (ad)is = $(20 \times 35)/110$; and for the last cell (cf) is = $(35 \times 50)/110$. After obtaining the expected values, the data are treated as in the first example.

Table 6-13. A 3×2 contingency table for a chi-square analysis of association between an isozyme locus (A) and a morphological trait locus (R).

	R^-		rr		
	O	E	O	E	Total
aa	15	18.56	8	6.19	
aa'	36	37.13	15	12.38	
a'a'	21	18.56	4	6.19	
Total	72		27		99

Table 6-14. A chi-square analysis of data where there is no known genetic ratio.

	a	b	c	Total
d	10	5	5	20
e	15	10	20	45
f	10	10	15	35
Total	35	25	50	110

6.9.2. *t*-Test of Differences Between Means

The conditions of a *t*-test are (*a*) random samples (n_1, n_2), (*b*) standard deviations of means are equal. The general formula for the statistic is

$$t = (\bar{x}_1 - \bar{x}_2)/s_p^2 \sqrt{(1/n_1 + 1/n_2)}.$$

s_p^2 = pooled variance and is obtained as

$$[(n_1 s_1^2 + n_2 s_2^2)/(n_1 + n_2) - 2],$$

where s_n^2 = sample variance. The degrees of freedom are = $n_1 + n_2 - 2$.

Example (hypothetical):

	n_1	n_2
	5	6
	6	10
	8	11
	6	9
	5	10
	6	14
	6	10
	7	9
	5	12
	5	10
Total	59	101
	$\bar{x}_1 = 5.9$	$\bar{x}_2 = 10.1$
	$n_1 = 10$	$n_2 = 10$

Sample variance is calculated as

$$s^2 = [(\Sigma x^2 - (\Sigma x)^2/n)]/n - 1,$$

where n = sample size. Where the second assumption of equal variance is violated, the variances cannot be pooled.

6.9.3. Standard Deviation

$s = \sqrt{s^2}.$

6.9.4. Coefficient of Variation

CV = $s/\bar{x}$. This statistic is usually expressed as a percentage.

Chapter 7

PRACTICAL APPLICATIONS OF ISOZYMES IN GENETIC ANALYSIS

The techniques of electrophoresis and isozyme analysis are used in a variety of ways too numerous to discuss separately. Hence this chapter discusses only a selected number of examples. One of the major applications of isozymes is as molecular markers. In this capacity, isozymes have advantages and disadvantages, some of which are discussed in this chapter. The methodologies presented here are slightly biased toward applications in plant genetic research, but the principles involved are applicable to both plant and animal studies. For example, I use the same buffer systems and staining protocols for both plant and fungal isozyme studies in common bean.

7.1. Isozymes as Genetic Markers

7.1.1. Advantages

The properties of isozymes confer upon them the following advantages over equivalent conventional techniques in research:

7.1.1.1. Genetic

1. Codominant gene expression. An advantage of codominance of gene expression is the ability of the researcher to detect heterozygote phenotypes (Crawford 1983; Rick and Tanksley 1983). For traditional markers, heterozygotes have to be classified together with one of the two classes of homozygotes because it is difficult or impossible to distinguish between them (i.e. these markers are usually under dominance-recessive gene expression).
2. Penetrance and expressivity. The genetic background in which a marker is placed may suppress its phenotype or influence its pattern of expression. When the phenotype is ascertained at the level of the protein (near the DNA), however, as is the case with isozymes, the masking effect of the genome is precluded.
3. Heritability. A trait has to be highly heritable to be used effectively as a marker. Some morphological markers such as some color traits are affected by the environment; such markers require more time and care when used in selection. Molecular markers, on the other hand, are completely heritable and hence decrease the chance of error in identifying the desired phenotype.
4. Marker pyramiding. Interlocus interaction (epistasis) is rare in isozymes and, unlike some morphological markers in which recessive alleles may produce deleterious effects in homozygotes (Moore and Collins 1983), molecular markers can, theoretically, be limitlessly assembled in an individual (Tanksley and Rick 1980) to produce a kind of "super genetic marker stock." The ability to produce an individual with multiple markers is attributable to the neutrality of isozymes, an advantage that has been exploited in the analysis of quantitative trait loci (QTLs).

7.1.1.2. Economic Whether the cost of using isozymes will be less than that of morphological markers will depend on several factors. An average biology laboratory will probably have some of the basic equipment needed for electrophoresis but not items such as power supply units, an electrophoresis unit, and reagents. The size of the study is also a factor in the cost of the technique. The issue of cost is not clear-cut.

7.1.1.3. Reliability Rick and Tanksley (1983) stated that a good electrophoretic result (zymogram) may permit the identification of positional differences as small as 1 mm between isozymes (bands). Isozyme analysis precludes many of the environmental effects that contribute to errors in identification and classification with morphological markers.

7.1.1.4. Practicality

1. Versatility. Nearly every plant tissue may be sampled for electrophoresis (Moore and Collins 1983). Hence the researcher has the option (sometimes) to choose the most convenient material (leaf, root, pollen, callus, seed, etc.). In animals, blood as well as solid tissue samples may be used in electrophoresis.

2. Flexibility. Isozyme work is conducted in a laboratory and hence is not subject to seasonal effects. Some conventional markers require screening of plants in a specific season, time of day, and stage in the life cycle of a plant. Screening by means of isozymes can be done out of season by planting in the greenhouse. Animal samples may be stored for use when convenient.

3. Breeders' rights protection. A breeder can breed a specific marker (or combinations of markers) into a cultivar for the purpose of providing an identification mark that will associate a breeder with a product (Pierce and Brewbaker 1973; Tanksley and Rick 1980).

7.1.1.5. Efficiency

1. Nondestructive sampling. Less than 200 mg of tissue is required for electrophoresis. This amount of tissue usually can be obtained from a plant without destroying it. In many crops with large seeds, it is possible to cut (or scrape) a piece of tissue from the raphe end of the seed and save the rest for planting. Leaves may be picked without damage to the plants and even roots may be sampled nondestructively. Blood samples may be taken from live animals in some cases without having to kill them.

2. Time and space. Breeding time and efforts can be drastically reduced if only plants with the desired genotypes are planted in the field. Planting both desirable and undesirable genotypes with the hope of successfully identifying the desirable types is the conventional breeding practice. Screening seedlings permits one to identify the desirable plants early in the breeding program instead of waiting for months or years as may be the case with perennial crops (Moore and Collins 1983). Furthermore, traditional plant breeding procedures frequently require testing of large numbers of segregating populations, a practice that requires a large space.

7.1.2. Disadvantages

7.1.2.1. Limited Number of Biochemical Markers The number of protocols available for isozyme work is rather limited. The proportion of the genome that such markers represent is rather small. Hence some researchers have looked for more diversity in their species by adopting more sophisticated procedures such as restriction fragment length polymorphisms (RFLP) and polymerase chain reaction (PCR), which have a higher chance of locating polymorphic loci in populations.

7.1.2.2. Safety Some chemicals used in electrophoresis are known or suspected carcinogens or are health hazards.

7.1.2.3. The Neutrality Issue Whereas in one sense the neutrality of isozymes has been found to be advantageous (Tanksley and Rick 1980), the same property is considered a source of bias in the sampling of allozymes (Brown and Clegg 1983). Most morphological genetic variation is adaptive.

7.1.2.4. Representativeness of Isozymes Protocols available for isozyme research assay enzymes that function especially in the glycolytic cycle, thereby making the technique biased and nonrandom with respect to certain studies. The reader may refer to Hubby and Lewontin (1966) for the criteria for genetic markers and the shortcomings of isozymes with respect to each criterion.

By highlighting the advantages of molecular markers, I do not in any way want to suggest that morphological markers are obsolete. Both markers will continue to be used in a complementary fashion.

7.2. Examples of the Applications of Isozymes in Research

7.2.1. Introduction

To list all the ways in which isozyme analysis can be used in solving research problems would be impossible. Limitation is imposed only by the imagination of the researcher. Isozymes are used in research basically by comparing patterns from different genotypes and looking for similarities and differences. This mode of application renders isozymes useful in basically three types of research:

1. *Time series* research of the progressive changes in genetic structure or levels of expression of genes.

2. *Cluster-divergence* studies that evaluate subjects on the basis of their similarity with each other.

3. *Import-export* research, which involves the deliberate movement of genes from one source to another.

Based on the above general categories of research, isozyme systems, if well-defined, can be used in various ways, including the following:

1. Research into the regulatory mechanisms of developmental processes characterized by differential expression of genes.
2. Research involving induced responses by the application of variable environmental factors (biotic and abiotic).
3. Systematic investigations.
4. Applications in "detective and policing" work (forensic, quality control).
5. Studies of population structure and dynamics.
6. Following gene transfers.

Because nearly every tissue from any level of cell organization in the organism may be sampled for isozyme determination, isozymes are applicable to cell and whole plant or animal research. Tissue culture researchers and breeders may apply electrophoresis to solve problems and facilitate their programs. As discussed above, breeders may capitalize on the attributes of molecular markers to increase their efficiency in selection.

7.2.2. Applications in Selected Fields of Plant Research

For a sampling of a few of the numerous, specific ways in which isozymes have been used in plant research, see the following references:

1. Isozymes in plant evolution: Rick et al. 1979; Gottlieb 1982; Doebley 1989.
2. Isozymes in population genetics: Arulsekar and Bringhurst 1981; Shaw and Allard 1982; Brown et al. 1989.
3. Isozymes and systematics: Cherry et al. 1970; Bringhurst 1981; Crawford 1989; Soltis and Soltis 1989.
4. Isozymes in tissue culture: Wetter and Kao 1976; Orton 1983.
5. Isozymes in the seed industry: Nijenhuis 1971; Tanksley and Jones 1982.
6. Isozymes as markers of quantitative trait loci: Tanksley et al. 1982; Stuber 1989.
7. Isozymes in plant-pathogen relationship studies: Micales et al. 1986.

7.2.3. Gene Mapping with Isozymes

One of the primary uses of isozymes is as biochemical markers of specific segments of chromosomes. Genes are arranged linearly on the chromosomes. Two gene pairs located on the same chromosome pair are said to be linked (Suzuki et al. 1981). Some gene pairs are so closely linked that they tend to be inherited *en bloc* as though they were a single gene. In genetic research, one may desire to trace the inheritance of specific traits. Similarly, in plant breeding, breeders often select genotypes with specific desirable traits from among segregating populations. In both cases, the desirable traits may be difficult or expensive to identify or select for directly. If the desirable traits are linked to known, simply inherited and readily selectable traits (markers), the traits may be indirectly selected for by selecting on the basis of the markers. Indirect selection is practiced also in animals.

The goal of gene mapping is to detect the specific locations of genes on chromosomes. The role of isozymes in this endeavor is ultimately to discover tight linkages between isozymic genes (as markers) and desirable traits (Tanksley 1982). Such a close association will allow a researcher readily to identify and select the linked desirable traits and thus facilitate a breeding program. Mapping is based on the principle that the closer two loci are together, the lesser the probability of a recombination occurring between them (Ayala and Kiger 1980). An example of such an association that has been put to practical use is the tight linkage between an acid phosphatase gene (*Aps-1*) and the nematode resistance gene (*Mi*), which has been successfully exploited in the transfer of nematode resistance from one cultivar to another (Rick 1983).

A linkage map shows the position of genes on a chromosome. Different kinds of markers (morphological, physiological, molecular, etc.) may be represented together on the same map. Linkage maps of some crops are relatively more saturated than others, for example corn (Goodman et al. 1980), tomato (Tanksley and Rick 1980), and wheat (Hart 1983). Different mapping techniques are in use. Three basic tasks are involved in gene mapping: (*a*) determination of the nature (tightness of linkage, conformation) of association between genes, (*b*) determination of the order of arrangement of genes on a chromosome, and (*c*) association of genes to specific chromosomes.

Some of the methods of gene mapping are briefly presented below.

7.2.3.1. Classical Linkage Tests One of the tests frequently conducted on isozyme data is deciphering the linkage relationships among enzyme-coding genes. This is accomplished by multipoint (two- or three-point) linkage tests that compare segregation of alleles at one locus with that of alleles at other loci for independent assortment (see worked examples in Sections 6.6 and 6.7). The major weaknesses of this classical test include the paucity of mapped morphological markers in most crop species and the large numbers of segregating populations required (Tanksley and Rick 1980). These weaknesses make the method a slow one to employ; however, the method may be made more efficient by capitalizing on some properties of isozymes (namely, codominance and lack of epistasis). In so doing, one can strategically choose parents to maximize the number of segregating isozyme loci per cross (Goodman and Stuber 1983).

7.2.3.2. Mapping with Aneuploids By using aneuploids, researchers are able to assign isozymes to specific chromosomes. One of the commonly used aneuploids is the *trisomic* (a diploid with an additional chromosome that is homologous to a pair of chromosomes in the normal chromosome complement). A trisomic series (i.e. each chromosome pair in the complement has an extra chromosome) should be available before this method can be used. Here, too, by capitalizing on the codominance of expression of most isozymic genes and the occurrence of allele dosage effect, researchers are able to undertake dosage mapping with isozymes. The disomic and trisomic zymogram phenotypes can be distinguished on the basis of the relative intensity of bands. In an F_1 progeny, the primary trisomic in which the isozyme gene in question occurs in triplicate will display a phenotype resulting from one dose of the allele from the male parent and two from the female trisomic parent. This will be different from the phenotype that results when the trisomic contributes one allele.

Dosage mapping with primary trisomics has been accomplished in several crops (Tanksley 1979). Different kinds of aneuploids as well as other chromosomal aberrations have been used in the mapping of isozyme genes. Most of the traits of interest to plant breeders are quantitatively inherited or under polygenic control. To use isozymes as markers to dissect quantitative traits, it is necessary to have many more mapped isozyme loci than is required for other applications. The reader is referred to Tanksley et al. (1982) and Stuber (1989) for applications of isozymes to analyzing quantitative trait loci.

7.2.4. Isozymes in Backcross Breeding

Backcrossing is a special breeding technique to transfer one or a few selectable genes from a donor to a recipient parent (Simmonds 1979). The goal of such a transfer is to enrich the recipient parent with the desirable donor genes without appreciable loss in genetic constitution. To achieve this end, the recipient parent is recurrently backcrossed to the F_1 hybrid and subsequent generations. The donor parent is frequently loaded with undesirable genes (apart from the target genes for which the breeding program was initiated), especially if it has wild origin. Since the backcross technique requires selection at each generation, the rate of progress in restoring the recipient parent (normally called *recurrent parent*) to its original or near-original genetic constitution depends on the selection efficiency, which in turn depends on the heritability of the traits used as selection criteria. Backcrosses are commonly conducted for many generations to achieve desirable end products. Any means by which the end results can be hastened is therefore very welcome.

Biochemical markers such as isozymes may be used to hasten a backcross breeding program by selecting for a set of isozymic marker loci scattered throughout the recipient genome. How this works is that a donor parent is strategically selected such that it differs from the recipient parent with respect to alleles at many isozymic loci. The loci, preferably, should be scattered throughout the genome to allow effective sampling of the genome. Selections from the first backcross are then subjected to isozyme analysis. The individuals that display all or most of the allozymes of the recipient parent are retained. The steps in a backcross breeding program using isozymes as proposed by Tanksley and Rick (1980) can be elaborated as follows:

1. Strategically choose a donor parent. This parent should differ from the recipient parent in as many isozyme alleles as possible.
2. Produce an F_1 hybrid.
3. Produce BC_1 progeny.
4. Select BC_1 individuals that display the traits (i.e. the donor traits) for which the backcross breeding program was initiated.
5. Evaluate selected BC_1 individuals to identify those which closely resemble the recipient parent.
6. Perform isozyme analysis on selected individuals from Step 5.
7. Evaluate Step 6 and retain those individuals that are homozygous for maximum number of recipient parent allozymes.
8. Produce BC_2 progeny by using individuals selected from Step 7.
9. Repeat from Step 4.

If an individual that displays the introgressed traits is identified at Step 7 and found to be homozygous at all the investigated isozyme loci, isozymes cease to be of further use in retrieving any recipient parental qualities that might remain through selections in subsequent backcross generations. The use of isozymes must be discontinued at that stage.

7.2.5. F_1 Hybrid Authentication

Sometimes contamination or improper crossing technique can cause the wrong pollen to be placed on a stigma, resulting in selfed seed instead of hybrid seed. A breeder may unknowingly advance a selfed seed and its progeny through several generations, thereby wasting time and resources because the desired improvement will not be realized. Markers are used in crossing programs to safeguard against this occurrence. When the parents involved in a cross are pure lines, the authentication of the F_1 seed is quite straightforward, insofar as the parents differ in the allele(s) used as marker(s) for the test. If one parent is fixed for one allele (AA) and the other parent is fixed for the alternative allele (aa), a true hybrid will have the genotype of Aa with respect to the A-locus. The property of codominance of expression of alleles at an isozymic locus will enable the three genotypes to be distinguished on the basis of their isozyme phenotypes (see examples of phenotypes in Chapter 5). An advantage of isozymes over morphological markers in hybrid authentication is that the test in most cases may be performed on immature F_1 plants.

When the parents involved in the cross are not true-breeding (i.e. one or both parents are segregating at the isozyme locus under investigation), hybrid authentication becomes more complex, requiring the estimation of allele frequencies (Arus 1983; Samaniengo and Arus 1983).

7.3. Isozymes in Fungal Research

Isozyme analysis is applicable to fungal genetic studies and has been used for taxonomic and genetic research by pathologists and mycologists. The staining protocols presented in this book have been used without modification in isozyme studies involving *Phaeoisariopsis griseola* (Sacc.) Ferr, the causal fungus of angular leaf spot (ALS) disease of common bean (*Phaseolus vulgaris* L.), in our laboratory at the Michigan State University. This section describes the procedure used in our laboratory for preparing fungal samples for electrophoresis.

7.3.1. Culturing Fungal Specimens

For organisms with large body mass, such as higher plants, obtaining adequate samples for electrophoresis may not be a problem. In fungi, however, culture of the pathogen is often necessary to increase the tissue mass and also to purify the tissue. Sufficient tissue type of certain fungi may be obtainable directly from the field, but subculturing may be necessary for some studies to ensure that only a single pathotype is analyzed.

The procedure for culturing may vary from one pathogen to another. The researcher should consult apppropriate sources for procedures suitable for a particular fungus. The procedure presented below was used by Correa-Victoria (1987) and Afanador (1989) for the angular leaf spot fungus. All isolates are purified by monospore transfer of cultures grown on V-8 commercial vegetable juice combined with agar according to the following recipe: V-8 juice = 125 ml, agar = 15 g, calcium carbonate = 2.6 g, and water = 1000 ml. The medium was stored at 4°C.

7.3.2. Types of Fungal Tissue Used as Samples

In higher plants, samples may be taken from nearly every part of the plant (leaf, seed, root, pollen, etc.). In fungi, samples may be taken from various vegetative and reproductive parts such as basidiocarps, ascocarps, conidia, mycelia, spores, and sclerotia (Micales et al. 1986). Fungi differ in the size of their fruiting bodies as well as other vegetative parts. The researcher may choose the most convenient for his/her purpose. It is critical that a uniform tissue sample be used because (*a*) the nuclear condition (mono- or dikaryotic) may differ from one tissue to another (Micales et al. 1986); (*b*) there is differential expression of alleles at different developmental stages (Hall 1967; Okunishi et al. 1979); and (*c*) the quantity of protein that can be obtained from sample preparation of the same tissue may differ (Bradford 1976).

To increase the chance of obtaining uniform samples, fungi samples should be collected at a specific stage of the growth curve of the fungus (Micales et al. 1986). The following procedure was used by Correa-Victoria (1987) and Afanador (1989) for angular leaf spot. Isolates are grown on V-8 medium in an incubator at 24°C for 10–14 days. Mycelial disks (4 mm diameter) are obtained from actively growing cultures. The disks are incubated at room temperature (24°C) on a shaker on 25 ml liquid modified Fries medium that contains the following: sucrose = 30 g, calcium.$2H_2O$ = 0.13 g, yeast extract = 10 g, water = 1000 ml. After 14 days, mycelia are harvested and vacuum-filtered on a nylon mesh. By rinsing with distilled water, the culture media is removed and the mycelial mats dried by blotting with sterile paper towel and stored at −20°C until needed.

7.3.3. Extraction of Fungal Protein

Micales et al. (1986) described different procedures for extracting proteins from various fungal tissues: (*a*) thin-walled tissue (e.g. some conidia) may be frozen in liquid nitrogen and crushed with a glass rod; (*b*) vegetative tissue and large fruiting bodies may be lyophilized to dehydrate them for easy crushing; (*c*) large fungal structures (e.g. mycelial mats, basidiocarps) may be homogenized in chilled mortar and pestle.

For *P. griseola*, Correa-Victoria (1987) adopted the following method: The extraction buffer consists of sucrose = 170 g, ascorbic acid = 1 g, cysteine HCl = 1 g, and 0.1 M tris-citrate buffer (pH 8.7) = 1000 ml. Five mg of the dry frozen mycelia is homogenized in 1 ml of extraction buffer in a mortar chilled on ice. To facilitate the grinding process, a small amount of acid-washed sand is added to the sample in the mortar. After grinding, the product is transferred into a 1.5 ml Eppendorf microcentrifuge tube and centrifuged at 2000 g for 20 min. The supernatant is then absorbed with a wick for electrophoresis as described above for plant material.

Authenticate isozymes whenever possible, even though sometimes the observed patterns can be compared to previously reported ones without much difficulty. To authenticate isozymes, one needs to be able to make crosses between parents carrying the different patterns of the expression of the enzyme. Although this is usually possible in most flowering plants, mating in fungal species is relatively more difficult to achieve. Afanador (1989) failed to obtain mating between sexual types of *P. griseola*.

Chapter 8

TROUBLESHOOTING

This chapter presents some of the common problems encountered in electrophoresis and their possible causes. For each problem, there may be other sources of error than stated. The researcher may discover, with practice, new ways of curbing certain problems. Remember to check the simpler possible sources of error before seeking complicated explanations.

Ultimately, the outcome of electrophoresis should be a zymogram that shows properly stained, resolved, and clearly separated isozyme bands for unambiguous scoring. Error at any of the stages of electrophoresis may sully the desired outcome. Pierce and Brewbaker (1973) stated that a successful electrophoresis is the cumulative result of a proper sample size, extraction buffer, gel composition, pH, electrolyte, electrical current, and staining technique. Standardizing the electrophoretic procedures for a laboratory is helpful so that problems may be more readily tracked down. Incorporating strategic checks into laboratory procedures (e.g. checking off chemicals on the staining protocol card as they are added to the staining solution) helps to reduce errors such as failure to add a staining solution component. Erradicating all sources of error is not easy. Some errors are certainly more costly than others. It is not uncommon to do everything right (at least you think so) and end up with poor results. Some sources of error can be identified quite readily and corrected before it is too late.

The problems included in this chapter have been categorized by the various stages of electrophoresis under which they are observed. By the time a problem is observed, however, it may be too late to correct it. Correction will have to be made in subsequent electrophoretic runs. The order of presentation under any category is arbitrary.

8.1. Troubleshooting in Starch Gel Electrophoresis

8.1.1. Buffers

1. PROBLEM: Buffer becomes mouldy.
   POSSIBLE CAUSE: Improper storage.
   CONSEQUENCE: Possibility of contaminating samples with foreign mould protein that may result in false isozymes (artifacts) on zymogram.
   SUGGESTED ACTION: (*a*) Store buffer in refrigerator, (*b*) prepare enough to last a short time.
2. PROBLEM: Buffer pH changes during storage.
   POSSIBLE CAUSE: Buffer was not stabilized during preparation (e.g. dry reagents used in the preparation of the buffer may not have completely dissolved).
   CONSEQUENCE: The quality of electrophoretic results for very pH-sensitive enzymes may change in later studies.
   SUGGESTED ACTION: Do not hurry the preparation of a buffer; take time to let pH stabilize before storing the buffer.

8.1.2. Sample Preparation

1. PROBLEM: Homogenate changes color (e.g. leaf samples change color from green to reddish brown) either in the mortar or on the wick.
   POSSIBLE CAUSE: Oxidation by phenolics, quinones, etc.
   CONSEQUENCE: Enzyme activity may be reduced (or even completely lost), resulting in poor staining results.
   SUGGESTED ACTION: (*a*) Add antioxidants and reducing agents (e.g. 2-mercaptoethanol, ascor-

bate) to extraction buffer; (*b*) try a different buffer (one with better protection against interfering substances and distabilization).

2. PROBLEM: Wick imbibes extract too slowly.
   POSSIBLE CAUSE: Homogenate may be too thick in consistency because of an insufficient amount of extraction buffer.
   CONSEQUENCE: (*a*) Wick will not absorb enough extract, and especially the top part of the wick may have no extract, so that the top slice of the gel will fail to stain for the enzyme under study; (*b*) too thick an extract may introduce too much protein into the gel (overloading), which may cause streaking upon staining.
   SUGGESTED ACTION: Use the proper ratio of buffer:tissue.

3. PROBLEM: Imbibed sample wicks have debris.
   POSSIBLE CAUSE: Unsatisfactory homogenization of solid sample.
   CONSEQUENCE: Insoluble materials may migrate into the gel to cause streaking upon staining.
   SUGGESTED ACTION: (*a*) Centrifuge sample after homogenization; (*b*) the wicks sometimes may be inverted at the time of loading so that the part that may have the debris is not in contact with the starch.

8.1.3. Starch Gel Preparation

1. PROBLEM: Lumpy, specky gel.
   POSSIBLE CAUSE: (*a*) Starch powder was not properly suspended (inadequate swirling) before hot buffer was added (in the case of the microwave oven preparation procedure) or while cooking over a Bunsen burner; (*b*) glassware and other containers used were dirty; (*c*) gel was degassed for too long a time so that the cooling and setting had started in parts of the gel when it was agitated.
   CONSEQUENCE: A non-uniform separation medium causes inaccurate mobility of proteins; specks may obstruct the migration of protein molecules in whose track they occur.
   SUGGESTED ACTION: (*a*) Clean all glassware and other containers used in starch preparation; (*b*) suspend the starch powder properly before adding the hot buffer; (*c*) do not agitate a gel that is in the process of setting in a mold; (*d*) if the number of specks are few, suck them out of the hot gel by using a Pasteur pipette.

2. PROBLEM: Starch gel is cloudy.
   POSSIBLE CAUSE: Insufficient cooking of the gel.
   CONSEQUENCE: Gel may have non-uniform consistency and may be prone to tearing during slicing.
   SUGGESTED ACTION: Cook for correct amount of the time next time.

3. PROBLEM: Starch gel is too thick and viscous.
   POSSIBLE CAUSE: (*a*) Insufficient amount of buffer; (*b*) prolonged degassing that leads to cooling of the gel while in the flask.
   CONSEQUENCE: (*a*) Not enough gel will be obtained to fill the mold to the appropriate volume, resulting in loss of a slice or two of gel (because some of the gel will remain on the walls of the flask); (*b*) starch concentration will increase leading to a slower electrophoretic run.
   SUGGESTED ACTION: (*a*) Reduce degassing time; (*b*) pour the gel quickly while it is still hot; (*c*) use correct starch:buffer ratio.

4. PROBLEM: Rough gel surface after cooling.
   POSSIBLE CAUSE: (*a*) Plastic cover on gel was not stretched smoothly over the surface (was wrinkled) during overnight storage; (*b*) the gel was wrapped before it was properly cooled, consequently trapping moisture underneath the cover to produce holes on the gel surface.
   CONSEQUENCE: (*a*) A thicker top slice may have to be discarded, with a possiblity of losing an otherwise useable slice; (*b*) non-uniform migration caused by obstructions to certain molecules (as a result of non-uniform flow of current) may arise.
   SUGGESTED ACTION: Always wrap the gel only after it is properly cooled and make sure the wrapper is stretched smoothly over the gel surface.

5. PROBLEM: Cracking noises heard when a gel is being poured into a mold.
   POSSIBLE CAUSE: Gel is too hot!
   CONSEQUENCE: Gel mold may crack!
   SUGGESTED ACTION: Let the gel cool just a little before pouring it into a mold. If degassing time is appropriate, no further cooling is desired.

8.1.4. Electrophoresis

1. PROBLEM: No current registers on the meter of the power supply unit.
   POSSIBLE CAUSE: (*a*) The cables may not be properly connected; (*b*) there may be a break in a connecting cable; (*c*) there may be improper contact due to worn sockets; (*d*) the power supply unit may not be plugged into mains or mains may be turned off.
   CONSEQUENCE: If the loaded gel is left unelectrophoresed for a long period, the proteins will start to diffuse in the gel and cause a blurred end-result upon staining.
   SUGGESTED ACTION: (*a*) Inspect power supply cables and clean contacts regularly; (*b*) check for proper connection of cables and setting of power supply unit.

2. PROBLEM: Marker dye moves too slowly.
   POSSIBLE CAUSE: (*a*) Low current—when connecting two gels to one power supply unit that

delivers constant current, the current dial must be set to two times the prescribed amount for one gel; a constant voltage power supply unit will automatically adjust the current to suit the number of gels); (*b*) gel and/or buffers may not have been properly prepared.
CONSEQUENCE: (*a*) Increased duration of electrophoresis; (*b*) the activity of certain enzymes that are sensitive to a prolonged period of electrophoresis will be jeopardized.
SUGGESTED ACTION: (*a*) Check power supply unit settings and correct; (*b*) change the buffers (use freshly prepared ones).

3. PROBLEM: Voltage too high.
POSSIBLE CAUSE: (*a*) The cloth wick (Type I electrophoresis model) may be too thin; (*b*) origin may not be properly closed; (*c*) buffers may not be of the right ionic strength; (*d*) in Type II models, the exposed gel in the legs of the mold may not be completely covered with electrolyte.
CONSEQUENCE: (*a*) Uneven protein migration and hence inaccurate relative mobilities; (*b*) poor resolution of bands; smearing.
SUGGESTED ACTION: (*a*) Close the origin properly after dewicking (e.g. insert a drinking straw between wall of the mold and the gel); (*b*) double the thickness of the absorbent cloth wick; (*c*) always add enough electrolyte to the tanks.

4. PROBLEM: Gel heats up.
POSSIBLE CAUSE: (*a*) Current too high; (*b*) poor cooling of the electrophoresis environment; (*c*) ionic strength of the buffers too high.
CONSEQUENCE: (*a*) Excessive heating of gel endangers especially the very temperature-sensitive enzymes; (*b*) gel slices may become brittle and tear easily during handling.
SUGGESTED ACTION: (*a*) Check the power supply settings periodically and adjust as necessary; (*b*) provide supplementary cooling (with water bag and ice block) when using buffer systems of high ionic strength.

5. PROBLEM: Recommended voltage and amperage difficult or impossible to attain without raising one of the electrical parameters to outrageously high level.
POSSIBLE CAUSE: (*a*) Buffers may not be properly prepared; (*b*) possibly a Type I-based buffer system is being run on a Type II system.
CONSEQUENCE: (*a*) Gel may heat up; (*b*) duration of electrophoresis may be prolonged, in which case the activity of some enzymes may be jeopardized.
SUGGESTED ACTION: (*a*) Change the buffer system; (*b*) use fresh buffer.

6. PROBLEM: Marker dye migrates in reverse direction.
POSSIBLE CAUSE: (*a*) Electrodes are reversed; (*b*) wrong dye was used.
CONSEQUENCE: If the gel strip on the side of the origin in which the migration is taking place is short, there is the risk of proteins migrating into the tank and getting lost.
SUGGESTED ACTION: (*a*) Reverse the electrodes (this action will delay the duration of electrophoresis); (*b*) check the marker dye to ensure that the proper one is being used.

8.1.5. Slicing of Gels

1. PROBLEM: Gel sticks to the bottom of the gel mold.
POSSIBLE CAUSE: (*a*) Dirty mold; (*b*) gel concentration is too low.
CONSEQUENCE: Bottom of the gel may break while it is pried from the mold, causing the bottom slice to be in danger of tearing while being lifted.
SUGGESTED ACTION: (*a*) Always ensure that the mold is very clean before pouring the starch gel into it; (*b*) run a wet, flat, flexible spatula across the bottom of the gel to free it from the mold.

2. PROBLEM: Gel is too hard to slice.
POSSIBLE CAUSE: (*a*) Gel may be too concentrated due to too much starch; (*b*) gel may be hardened by excessive cooling during electrophoresis; (*c*) slicer string may be loose or too thick.
CONSEQUENCE: (*a*) The gel slices may have rough surfaces that could cause poor results upon staining; (*b*) gels may be brittle and tear easily during handling.
SUGGESTED ACTION: (*a*) Reduce starch concentration; (*b*) use a finer wire [guitar E-string (6th)] for slicing; (*c*) adjust the cooling environmental temperature to about 4°C.

3. PROBLEM: Gel slices have rough surfaces.
POSSIBLE CAUSE: (*a*) Slicer string may be dirty; (*b*) string was not drawn smoothly through the gel.
CONSEQUENCE: (*a*) Weak spots may develop in the gel slices and predispose them to tearing; (*b*) the stained gel may show differential staining intensities because some bands are located in very thin spots with little enzyme.
SUGGESTED ACTION: Wipe or wash the slicing string between slicings with a damp cloth.

8.1.6. Staining

1. PROBLEM: Dye front is wavy.
POSSIBLE CAUSE: (*a*) Cloth wick front may be uneven; (*b*) slit (origin) may not have been straight and smoothly cut; (*c*) presence of air pockets in the slit.
CONSEQUENCE: (*a*) Relative mobility of proteins will not be accurate for comparison; (*b*) for certain studies, the results will not be of any use.
SUGGESTED ACTION: (*a*) Always make a clean and straight cut at the origin for the sample wicks and

ensure that the cloth wick is set up so that its edge is straight and parallel to the origin; (*b*) close up the slit after dewicking.

2. PROBLEM: Dye front is straight but slanted.
POSSIBLE CAUSE: (*a*) The gel may be of uneven thickness; (*b*) cloth wick may not be parallel to the origin; (*c*) origin may have opened up on one side.
CONSEQUENCE: Scoring proteins whose relative mobility values differ by a few milimeters may be difficult.
SUGGESTED ACTION: (*a*) Always place the gel mold on a level surface before pouring the starch gel; (*b*) the absorbent cloth wick should have a straight edge set parallel to the origin.

3. PROBLEM: Dye front is curved (smiling or sad face zymogram).
POSSIBLE CAUSE: (*a*) Non-uniform cooling of gel (if the center is colder than the edges, as might happen if an ice block is used as a supplemental cooling aid, the gel front will curve up at the edges of the gel); (*b*) improper closing of the origin at the edges (opening of the origin at the edges will retard the migration of the proteins in those areas).
CONSEQUENCE: Scoring may be complicated. Adjacent bands at the edges may merge and mobility may be accelerated or retarded in those regions of the gel.
SUGGESTED ACTION: (*a*) Always cover the exposed parts of the gel with a plastic wrap since the cooling environment has drying effects that cause the gel to dry and shrink and the origin to open; (*b*) inserting a drinking straw (as previously described) will help in keeping the slit at the origin closed during electrophoresis.

4. PROBLEM: Adjacent bands merged.
POSSIBLE CAUSE: (*a*) Sample wicks were loaded too close together; (*b*) some wicks were slanted in the gel at the time of loading; (*c*) excess stain was not blotted away from the wicks before loading; (*d*) too much sample extract was introduced into the gel; (*d*) gel was not electrophoresed soon after loading or was not stained soon after electrophoresis, thus allowing some diffusion of proteins in the gel.
CONSEQUENCE: Scoring will be problematic and sometimes impossible.
SUGGESTED ACTION: (*a*) Space wicks properly; (*b*) use narrower wicks to reduce amount of extract loaded; (*c*) blot away excess extract from wicks before loading; (*d*) keep wicks erect in the slit at the origin.

5. PROBLEM: A greater number of clearly stained lanes than the number of lanes (number of samples) loaded.
POSSIBLE CAUSE: (*a*) Rearrangement of wicks in the starch gel (this can happen if one is not using a loading guide and finds that several wicks are left over with no place on the gel to place them); (*b*) the wick with the marker dye tends to absorb proteins from the adjacent wicks if placed close to a sample wick (especially to one that is saturated with sample extract), so that an extra lane usually identical to the adjacent one in pattern is observed; (*c*) wicks were not rid of excess sample extract, which consequently spread into spaces between adjacent wicks.
CONSEQUENCE: Confusion at the time of scoring may lead to erroneous scoring.
SUGGESTED ACTION: (*a*) Use a loading guide, especially if one is a neophyte at the technique of electrophoresis; (*b*) avoid shifting the wicks once they are inserted in the origin slot, especially if they are saturated with extract; (*c*) use a very thin wick for the marker dye and place it a good distance from the nearest specimen wick; (*d*) place wicks on absorbent material to blot out excess sample extract before loading a gel.

6. PROBLEM: Bands are clearly resolved but closer than normal.
POSSIBLE CAUSE: Insufficient duration of electrophoresis.
CONSEQUENCE: Scoring may be a problem in some cases.
SUGGESTED ACTION: Stick to the recommended duration of electrophoresis.

7. PROBLEM: Bands are poorly resolved and separated.
POSSIBLE CAUSE: (*a*) Buffer system not suitable for the enzyme; (*b*) pH and ionic strength of buffer may not be correct; (*c*) too much sample extract introduced into the gel; (*d*) diffusion of proteins in the gel.
CONSEQUENCE: Scoring may be problematic or even impossible.
SUGGESTED ACTION: (*a*) Use a different buffer system for the particular enzyme; (*b*) recheck the pH of the buffers (and also the calibration of the pH meter); (*c*) use smaller wicks and dewick the gel after 15–20 minutes of electrophoresis ("pre-electrophoresis"); (*d*) try an overlay technique of staining for the enzyme.

8. PROBLEM: Streaks or smears occur between bands.
POSSIBLE CAUSE: (*a*) Inappropriate buffer system; (*b*) too much sample extract has been loaded for electrophoresis; (*c*) unsuitable slice of gel was used for staining.
CONSEQUENCE: Scoring and comparison will be complicated.
SUGGESTED ACTION: (*a*) Change the buffer system or check the pH as mentioned above (a buffer of

higher ionic strength tends to give a sharper resolution than one of a lower ionic strength); (*b*) take steps to reduce the amount of extract electrophoresed (e.g. dewick, use smaller wicks, blot away excess extract); (*c*) reshuffle the staining order (i.e. try a gel slice from another location in the stack of slices).

9. PROBLEM: Differential staining intensity of bands.
   POSSIBLE CAUSE: (*a*) Differences in the physiological state, handling, and preparation of the samples prior to electrophoresis; (*b*) different concentrations of sample extracts loaded (e.g. due to differences in the amounts of extraction buffer soaked up by wick, wick sizes).
   CONSEQUENCE: Scoring becomes complicated and results are prone to misinterpretation.
   SUGGESTED ACTION: Standardize the electrophoretic procedure so that samples of the same tissue type, age, etc. are used and treated alike.

10. PROBLEM: Bands are too faint to score.
    POSSIBLE CAUSE: (*a*) Unsuitable buffer system; (*b*) pH may be unsuitable; (*c*) staining solution may not have been well-prepared (e.g. reduced amounts of certain components); (*d*) staining solution chemicals not in good condition or not the proper grade; (*e*) enzyme activity may be weak due to deterioration or insufficient amount; (*f*) insufficient incubation time.
    CONSEQUENCE: Bands may be unscoreable.
    SUGGESTED ACTION: (*a*) Use the proper grades of chemicals and do not substitute with inferior ones; (*b*) ensure that the staining solution is complete; (*c*) store all chemicals properly to keep them in good condition; (*d*) try other recipes for staining; (*e*) prolong the staining time; (*f*) use a gel slice of double the usual thickness.

11. PROBLEM: Bands are too darkly or deeply stained.
    POSSIBLE CAUSE: (*a*) Too much protein extract electrophoresed; (*b*) overstaining.
    CONSEQUENCE: Bands that are close together may merge and cause scoring complications.
    SUGGESTED ACTION: (*a*) Discontinue the staining process at the correct time; (*b*) if scoring will not be done immediately, fix the gel appropriately.

12. PROBLEM: Background of stained gel is too dark.
    POSSIBLE CAUSE: Exposure of light-sensitive stains to light during the staining process (at preparation of staining solution or incubation).
    CONSEQUENCE: In some cases, lighter-staining bands may be obscured and missed during scoring.
    SUGGESTED ACTION: (*a*) Incubate in the dark unless otherwise indicated; (*b*) store all stains in the dark.

13. PROBLEM: Staining solution changes to unusual color while being prepared.
    POSSIBLE CAUSE: (*a*) Premature formation of formazan; (*b*) wrong timing in adding certain components to the solution.
    CONSEQUENCE: Frequently, the deteriorated staining solution will not stain a gel properly.
    SUGGESTED ACTION: Follow the recommendations of the recipe.

14. PROBLEM: Gel is "blank" (i.e. no bands where normally there are bands).
    POSSIBLE CAUSE: (*a*) Staining solution may be missing at least one component; (*b*) samples may have deteriorated (due to improper extraction process, poor storage); (*c*) poor condition of the substrate, enzyme, or one of the staining solution components; (*d*) improper grade of a staining component item (e.g. one tries to substitute a recommended item with a new one or use an NADP-active enzyme where an NAD one is required).
    CONSEQUENCE: Wasted effort; nothing to score.
    SUGGESTED ACTION: (*a*) Prepare staining solution more carefully; (*b*) use fresh samples and new stock of chemicals; (*c*) check the coenzyme where applicable; (*d*) use the recommended grade and amounts of staining solution components.

15. PROBLEM: Irregular number of bands.
    POSSIBLE CAUSE: (*a*) Different age (developmental) of samples; (*b*) different physiological conditions of samples; (*c*) protein undergoing changes in quaternary structure as a result of handling conditions; (*d*) stain artifacts (certain parts of protein molecules interacting with the staining solution).
    CONSEQUENCE: Inconsistent results and complications when interpreting zymograms may lead to wrong conclusions.
    SUGGESTED ACTION: (*a*) Use fresh or properly stored samples; (*b*) use more than one buffer system to confirm results and authenticate isozymes; (*c*) when applicable, use samples at the same developmental stage.

8.1.7. Fixing Gels

1. PROBLEM: Bands fade quickly even after fixing.
   POSSIBLE CAUSE: Inappropriate fixative.
   CONSEQUENCE: Records will be lost unless the gel was scored prior to fixing.
   SUGGESTED ACTION: Try another fixative.

8.2. Troubleshooting in Polyacrylamide Gel Electrophoresis

Whereas some of the problems discussed under SGE apply to PAGE, others are peculiar to PAGE. Some of these are as follows:

8.2.1. Polyacrylamide Gel Preparation

1. PROBLEM: Polymerization rate is too fast.
   POSSIBLE CAUSE: Concentration of catalyst is too high.
   CONSEQUENCE: Polymerization may occur before one has the chance to pour the gel mixture into the mold.
   SUGGESTED ACTION: Use a weaker concentration of catalyst next time.
2. PROBLEM: Polymerization is too slow.
   POSSIBLE CAUSE: Concentration of catalyst is too weak (acidic gels generally require a longer time for polymerization).
   CONSEQUENCE: Delay of work.
   SUGGESTED ACTION: *(a)* Use more of the same catalyst; *(b)* use a more efficient catalyst.
3. PROBLEM: Slanted, uneven gel surface.
   POSSIBLE CAUSE: *(a)* Gel mold was not set on a level surface prior to pouring the gel; *(b)* water layer was not properly floated on top of the polymerizing gel mixture, thereby causing agitation.
   CONSEQUENCE: *(a)* Gel cannot be reliably used, especially for separating proteins with unfamiliar banding patterns; *(b)* other determinations such as molecular weight will not be accurate.
   SUGGESTED ACTION: Always ensure that the casting stand or gel mold is properly leveled prior to casting the gel mixture.

8.2.2. Loading Samples

PROBLEM: Sample fails to stay at the bottom of the well (i.e. diffuses into the buffer solution).
POSSIBLE CAUSE: Weak "anchor" or absence of it.
CONSEQUENCE: Contamination of samples.
SUGGESTED ACTION: *(a)* Use the proper anchor; *(b)* when loading, do not push on the pin of the syringe until the needle is inserted in the well.

8.2.3. Electrophoresis

PROBLEM: Samples migrate in an unexpected direction (e.g. up into the upper buffer).
POSSIBLE CAUSE: Improper electrical connections.
CONSEQUENCE: If the problem is not found and corrected in time, samples will be contaminated and data will eventually be lost.
SUGGESTED ACTION: *(a)* Check the electrode connections and properly connect cords; *(b)* bear in mind that certain procedures require the electrode connections to be *reversed*.

8.2.4. Unloading a gel

PROBLEM: Gel sticks to the plate or tube.
POSSIBLE CAUSE: *(a)* Dirty plate or tube; *(b)* low concentration of gel.
CONSEQUENCE: Gel may be torn or broken during handling.
SUGGESTED ACTION: Clean plates very thoroughly before use.

8.2.5. Staining Gels

1. PROBLEM: Bands found in lanes originally not loaded with samples.
   POSSIBLE CAUSE: The upper buffer was contaminated during loading of samples. This may happen when additional buffer is added rapidly to the upper reservoir so that it agitates the solution.
   CONSEQUENCE: *(a)* Dark background of stained gels; *(b)* complications in scoring of gel; *(c)* gel may not be scoreable.
   SUGGESTED ACTION: Load gels carefully by pushing down the pin of the syringe very gently and only when the needle is in the well; empty the syringe completely before withdrawing the needle.
2. PROBLEM: Streaking of stained gel.
   POSSIBLE CAUSE: Nucleic acids become precipitated by ampholines.
   CONSEQUENCE: Large spots appear instead of sharp bands and hence resolution is decreased.
   SUGGESTED ACTION: *(a)* Add DNase and RNase (O'Farrell 1975); *(b)* centrifuge samples prior to using.

Chapter 9

ORGANIZING A LABORATORY FOR ELECTROPHORETIC RESEARCH

In this chapter, a list of equipment and supplies and a sample list of suppliers are provided. The lists are by no means exhaustive of all the items one would ever need for electrophoresis. One may find it necessary to substitute other items during the course of research or find other suppliers. The lists provide only a start for setting up a laboratory for electrophoresis following the procedures outlined in this book. The mention of a trademark, a specific type of equipment, or proprietary product should not be construed as an endorsement.

9.1. General Laboratory Equipment

The following general equipment may be available in an average biology laboratory:

1. Refrigerator
2. Freezer
3. Porcelain spot plates and pestles
4. pH meter
5. Flat top balance (should be able to weigh amounts less than 10 mg)
6. Analytical balance
7. Incubator/oven
8. Spatulas (different sizes)
9. Magnetic stirring bars (different sizes)
10. Measuring cylinders (different volumes from 10 ml to 1 liter)
11. Containers for buffer solutions (different sizes from 100 ml to 6 liters)
12. Funnels (different sizes)
13. Erlenmeyer flasks (different sizes)
14. Filter flask (1 liter)
15. Volumetric flask (1 liter)
16. Beakers (different sizes)
17. Ruler
18. Pair of scissors
19. Scapel/knife
20. Wash bottles
21. Pipettes (different sizes)
22. Whatman paper (Nos. 1, 2, 3)
23. Absorbent paper towel
24. Thermometers
25. Dessicant
26. Fume hood
27. Disposable latex examination gloves
28. Forceps (blunt tip)

9.2. Additional Equipment

The type and amount of additional equipment required specifically for electrophoresis will depend on the electrophoretic method (SGE or PAGE) to be used. Some of the special equipment is listed below:

1. Microcentrifuge
2. Staining trays (e.g. pyrex dishes, tupperware dishes)
3. Microcentrifuge tubes and racks
4. Plastic wrap
5. Suction system (aspirator); tap model
6. Ice bucket
7. Pipette dispenser and tips
8. Water bags
9. Reuseable ice packs
10. Source of ultraviolet light
11. Source of DC power supply
12. Electrophoresis set
13. Motor unit for motorized grinding
14. Light table

9.3. Chemicals

The chemicals in the following lists have been classified according to general use.

9.3.1. Electrophoresis Media for Starch Gel Electrophoresis

1. Starch (potato, hydrolyzed)
2. Connaught starch
3. Electrostarch

9.3.2. Electrophoresis Media for Polyacrylamide Gel Electrophoresis

1. Acrylamide
2. *N,N'*-Methylene-bis-acrylamide (bis)
3. Agarose

9.3.3. Buffer Chemicals

1. *N*-3-Aminopropyl morpholine
2. L-Ascorbic acid (sodium salt)
3. L-Ascorbic acid
4. Arsenic acid (sodium salt)
5. Boric acid
6. Citric acid (monohydrate)
7. Citric acid (anhydrous)
8. Glycine (sodium salt)
9. L-Histidine
10. DL-Histidine
11. Hydrochloric acid
12. Lithium hydroxide
13. Maleic acid
14. Potassium phosphate (monobasic salt)
15. Potassium phosphate (dibasic salt)
16. Potassium hydroxide
17. Sodium acetate
18. Sodium phosphate (monobasic salt)
19. Sodium phosphate (dibasic salt)
20. Sodium hydroxide
21. Tris (Trizma base)
22. Urea

9.3.4. Stains/Dyes

1. Amido black
2. Bromophenol blue
3. Coomassie brilliant blue-250
4. *o*-Diansidine, tetrazotized (Diazo blue B)
5. 2,6-Dichlorophenol-indophenol (DCPP)
6. Fast blue BB salt
7. Fast blue RR salt
8. Fast black K salt
9. Fast garnet GBC salt
10. Fast green
11. Ferric chloride
12. Meldola's blue
13. Methylene blue
14. Methyl green
15. Nitroblue tetrazolium (NBT)
16. Naphthol blue black
17. Phenazine methasulphate (PMS)
18. Potassium iodide
19. Potassium ferricyanide
20. Thiazolyl blue tetrazolium (MTT)

9.3.5. Anti-oxidants and Miscellaneous Buffer Additives

1. Dithiothreitol (DTT)
2. Ethylenediamine tetra-acetic acid (EDTA) (disodium salt)
3. 2-Mercaptoethanol
4. Polyvinylpyrrolidone (PVP–40)
5. Sucrose
6. Triton X-100

9.3.6. Coenzymes

1. β-Nicotinamide adenine dinucleotide (NAD)
2. β-Nicotinamide adenine dinuclotide phosphate (NADP) (sodium salt)
3. β-Nicotinamide adenine dinucleotide reduced (NADH)
4. Peridoxal-5-phosphate
5. Riboflavin

9.3.7. Cofactors

1. Magnesium chloride
2. Manganese chloride
3. Magnesium sulfate
4. Calcium chloride

9.3.8. Natural Substrates

1. *cis*-Aconitic acid
2. Aconitic anhydride
3. Adenosine 5′-diphosphate (ADP)
4. Adenosine 5′-triphosphate (ATP)
5. L-Arginine β-naphthylamide
6. L-Aspartic acid
7. D-Fructose-6-phosphate (disodium salt)
8. β-D(+)-Glucose
9. α-D-Glucose-1-phosphate (disodium salt)
10. L-Glutamic acid
11. DL-Glyceric acid
12. β-Glycerophosphate
13. DL-Isocitric acid (trisodium salt)
14. α-Ketoglutaric acid
15. DL-Malic acid
16. L-Malic acid
17. 6-Phosphogluconic acid (trisodium salt)
18. Shikimic acid
19. Sorbitol

9.3.9. Artificial Substrates

1. 6-Bromo-2-naphthyl-β-D-glucopyranoside
2. Hydrogen peroxide (30% solution)
3. α-*N*-Benzoyl-DL-arginine-β-naphthylamide HCl (BANA)
4. L-Leucyl β-naphthylamide
5. Menadione
6. 4-Methylumbelliferyl-β-D-glucoside
7. 4-Methylumbelliferyl-*N*-acetyl-β-D-glucosamine

8. α-Naphthyl acetate
9. β-Naphthyl acetate
10. α-Naphthyl acid phosphate (sodium salt)
11. Dihydroxyacetone phosphate (DHAP)

9.3.10. Enzymes

1. Glyceraldehyde-3-phosphate dehydrogenase
2. β-Glycerophosphate dehydrogenase
3. Glucose-6-phosphate dehydrogenase
4. Hexokinase
5. Isocitric dehydrogenase

9.3.11. Solvents

1. Acetone
2. *N,N'*-Dimethyl formamide
3. Ethanol (95%)

9.3.12. Fixation and Storage Chemicals

1. Acetic acid (glacial)
2. Glycerol
3. Methanol
4. Trichloroacetic acid (TCA)

9.3.13. Special Chemicals for Polyacrylamide Gel Electrophoresis

1. Agar
2. Ampholines
3. β-Alanine
4. Ammonium persulfate
5. *N,N,N',N'*-tetramethyl-ethlyenediamine (TEMED)
6. Sodium dodecyl sulfate (SDS)
7. Ferrous sulfate
8. Molecular weight markers

9.4. Selected Suppliers

A list of suppliers is provided to help the beginner to get started in setting up a laboratory for electrophoresis. It is not comprehensive, nor should be construed as an endorsement. The suppliers are located in North America and may have subsidiaries in various parts of North America and other parts of the world.

1. Aldrich Chemical Company, Inc., 940 West Saint Paul Avenue, Milwaukee, WI 53233 CHEMICALS, STARCH
2. American Scientific Products, 1430 Wankegan Road, McGaw Park, Illinois 60085-6787 CHEMICALS
3. Connaught Laboratories, Ltd., P. O. Box 1755, Station A, Willowdale, Ontario, Canada M2N 5B8 STARCH
4. Connaught Laboratories, Ltd., Route 611, Swift Water, PA STARCH
5. E-C Apparatus Corporation, 3831 Tyrone Boulevard N., St. Petersburg, Florida 33709 POWER UNIT
6. Fisher Scientific Co., Ltd., 112 Colonnade Rd., Nepean, Ontario, Canada K2E 7L6 CHEMICALS, STARCH
7. Fisher Scientific Co., Ltd., 52 Fadem Road, Springfield, NJ 07081 CHEMICALS, STARCH
8. Heath Company, Benton Harbor, MI 49022 POWER UNIT
9. LKB Instruments, Inc., 12221 Parklawn Drive, Rockville, MD 20852 POWER UNIT
10. Sigma Chemical Corporation, P. O. Box 14508, St. Louis, MO 63178 CHEMICALS
11. US Biochemical Corporation, P. O. Box 22400, Cleveland, OH 44122 CHEMICALS

9.5. Storage of Chemicals

Chemical orders are usually accompanied by storage and handling instructions on their containers. It is good practice to write the date of purchase of each item on the container. Some chemicals require no special storage conditions and may be simply stored in a cool dry place. Some require freezer storage with a dessicant while others require storage in a refrigerator (0–5°C) or freezer (below 0°C). One may consult the Merck™ index for information pertaining to health hazards of any chemical used for electrophoresis or other laboratory work. Some chemical companies accompany their orders with some information on the safe handling and use of their products.

9.6. Useful Conversions

TEMPERATURE: °F = (9/5)C + 32, °C = (5/9)F—32
WEIGHTS: 1 g = 1000 mg, 1 mg = 0.001 g
VOLUME: 1 ml = 1000 μl
CONCENTRATION: 1 mM = 0.001 M
LENGTH: 1 cm = 0.39 in

Chapter 10

STAINING PROCEDURES FOR SELECTED ENZYMES

10.1. Introduction

This chapter provides staining protocols for enzymes frequently assayed in plants and animals. The protocols differ in cost and ease of application and safety. I do not claim credit for developing any of these protocols, although I did modify some for better results. As much as possible, the original author(s) of each procedure has been acknowledged. Published protocols are sometimes modifications of original protocols, because a recipe may have been developed for one species and later adapted to others. Many of the recipes used in plants were adapted from protocols for animal research. When multiple references are cited for one protocol, the first named is the original author, and the next is the researcher(s) who modified it. The older references are usually recipes from animal studies.

Each procedure as listed here is preceded by the enzyme trivial name, a common abbreviated name (usually a three-letter acronym), and its Enzyme Commission (E.C.) number. The E.C. number is usually a four-section number in which sections are separated by periods. The sections of the E.C. number correspond to the main class, subclass, sub-subclass, and serial number of the enzyme, respectively.

The quaternary structure and a brief account of the physiological function (where available) of each enzyme or some other special characteristic are also given. The chemical basis for each staining procedure is presented in the form of a reaction(s) to show the steps involved in obtaining a chromatic effect. Most of the staining schedules utilize the tetrazolium system for producing a chromatic effect. Diagrams of some of the major metabolic pathways illustrate the position of an enzyme in a metabolic pathway or cycle. Since the pH of buffers is frequently manipulated, alternative protocols are included with new methodology or different reagents.

Most protocols require up to an hour to complete the staining process but may sometimes take longer than recommended to adequately observe stained bands. The staining process must be terminated at the right time for reliable scoring or effective fixing of gels because some enzymes are sensitive to overstaining. Prolonged staining may introduce secondary bands that can complicate the interpretation of electrophoretic results and model-building.

Most of the enzymes included in this chapter can be stained at room temperature, but incubating at 37°C is generally recommended for good, rapid results. All the protocols, irrespective of their source, have been standardized for preparing 50 ml of staining solution unless otherwise indicated. When water is specified in a recipe, distilled water should be used. The reaction formulae were modified after Brewer and Sing (1970) and Vallejos (1983).

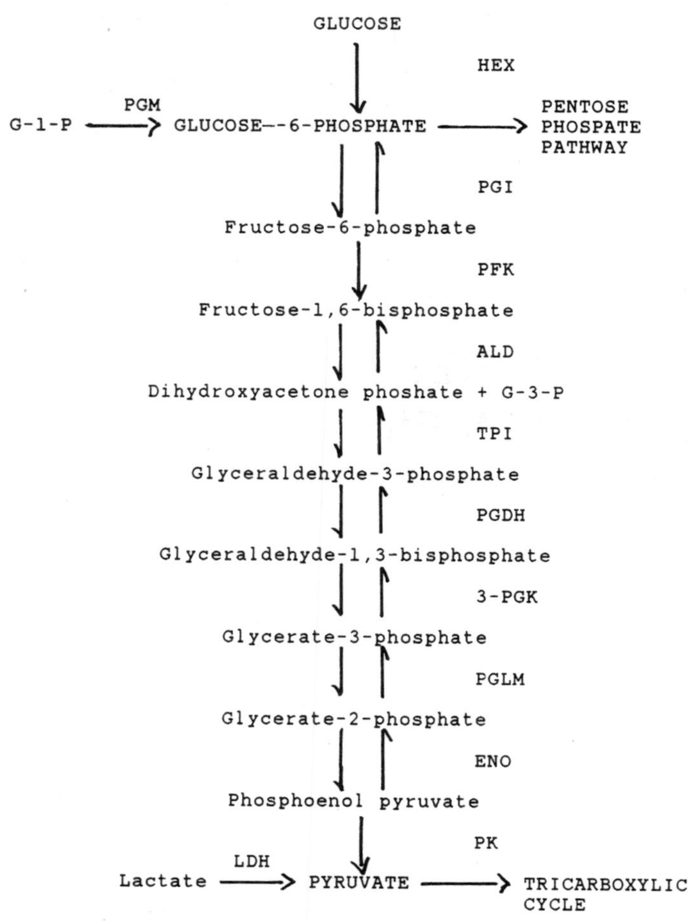

Figure 10-1. Enzymes of the glycolytic pathway.
LEGEND: HEX = hexokinase, PGM = phosphoglucomutase, PGI = phosphoglucose isomerase, PFK = phosphofructokinase, ALD = aldolase, TPI = triosephosphate isomerase, PGDH = phosphoglyceraldehyde dehydrogenase, 3-PKG = phosphoglycerate kinase, PGLM = phosphoglyceromutase, ENO = enolase, LDH = lactate dehydrogenase, PK = pyruvate kinase.

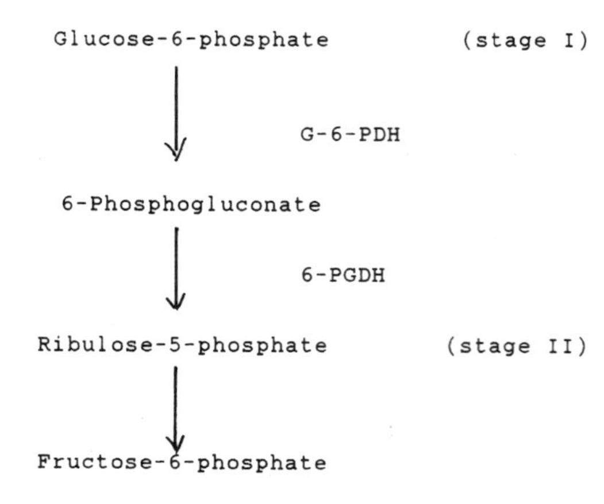

Figure 10-2. Enzymes of the pentose phosphate pathway.
LEGEND: G-6-PDH = glucose-6-phosphate dehydrogenase, 6-PGDH = phosphogluconate dehydrogenase.

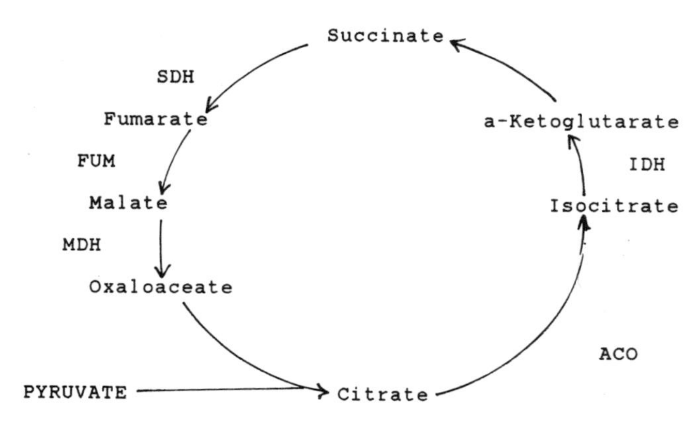

Figure 10-3. Enzymes of the tricarboxylic acid pathway.
LEGEND: SDH = succinate, FUM = fumarase, ACO = aconitase, MDH = malate dehydrogenase, IDH = isocitrate dehydrogenase.

10.2. Classes of Enzymes

Zubay (1983) recognizes six main classes of enzymes: *oxidoreductases, transferases, hydrolases, lyases, isomerases,* and *ligases* (or *synthases*). These six classes are sometimes further broken down (e.g. oxidoreductases may be subdivided into *dehydrogenases* and *oxidases*). The list of protocols presented in this book includes a wide range of enzyme classes for which suitable assays have been developed.

10.2.1. Oxidoreductases

The enzymes in the oxidoreductase class catalyze oxidation-reduction reactions. Dehydrogenases catalyze the dehydrogenation of substrates by using non-oxygen molecules as hydrogen acceptors, and oxidases catalyze the oxidation of substrates by using oxygen molecules as electron acceptors (Mahler and Cordes 1968).

10.2.1.1. Dehydrogenases

1. Alcohol dehydrogenase (ADH)
2. Glucose-6-phosphate dehydrogenase (G-6PD)
3. Glutamate dehydrogenase (GDH)
4. Glycerate-2-dehydrogenase (G2D)
5. Isocitric dehydrogenase (IDH)
6. Malate dehydrogenase (MDH)
7. Malic enzyme (ME)
8. 6-Phosphogluconate dehydrogenase (6PGD)
9. Shikimate dehydrogenase (SKDH)
10. Sorbitol dehydrogenase (SODH)
11. Succinate dehydrogenase (SDH)
12. Lactate dehydrogenase (LDH)
13. Xanthine dehydrogenase (XDH)

10.2.1.2. Oxidases

1. Aconitase (ACO)
2. Catalase (CAT)
3. Diaphorase (DIA)
4. Superoxide dismutase (SOD)
5. Peroxidase (PRX)

10.2.2. Transferases

Transferases catalyze the transfer of functional groups between two substrates.

1. Adenylate kinase (ADK)
2. Aspartate aminotransferase (AAT or GOT)
3. Hexokinase (HK)
4. Phosphoglucomutase (PGM)

10.2.3. Hydrolases

Hydrolases catalyze reactions involving hydrolysis.

1. Acid phosphatase (ACP)
2. Alkaline phosphatase (AP)
3. Endopetidase (ENP)
4. Fluorescent esterase (FLE)
5. Fructose-6-diphosphatase (F6DP)
6. β-D-Glucosidase (GLU)
7. Leucine aminopeptidase (LAP)
8. *N*-Acetyl glucoseaminidase (NAG)

10.2.4. Lyases

Lyases catalyze reactions that add units to double bonds.

1. Aldolase (ALD)
2. Fumarase (FUM)

10.2.5. Isomerases

Isomerases catalyze isomerization reactions.

1. Phosphoglucose isomerase (PGI)
2. Mannose phosphate isomerase (MPI)
3. Triosephosphate isomerase (TPI)

10.2.6. General Proteins

General proteins are not enzymes because they have no catalytic activity. They are found in large quantities in an organ or organism. Some examples are as follows:

1. Phaseolin (PHA), found in dry bean
2. Ribulose-bisphosphate carboxylase (Rubisco) (RBCO), found in plant leaf
3. Hemoglobin (HEM), found in animals.

10.3. Biochemical Pathways

Staining protocols have been developed for some enzymes that function in glycolysis, gluconeogenesis, pentose phosphate pathways, and the tricarboxylic acid cycle. The biochemical pathways and the enzymes involved are presented in Figures 10-1 to 10-3.

10.4. Protocols for Histochemical Staining of Enzymes

The protocols for histochemical staining of enzymes are presented in alphabetical order.

10.4.1. β-*N*-Acetylglucoseaminidase (NAG) E.C. 3.2.1.30

Staining produces a fluorescent product that is observed under a u.v. light.

Example

BUFFER: 0.1 M Na citrate, pH 4.5 = 40 ml
4-Methylumbelliferyl-*N*-acetyl-β-D-glucosaminide = 25 mg
(dissolve in 10 ml acetone)
INSTRUCTIONS: Incubate the gel at room temperature for 8 min. Rinse and score under longwave u.v. light.
SOURCE: Weeden and Emmo (n.d.).

10.4.2. Acid Phosphatase (ACP) E.C. 3.1.3.2

Acid phosphatases (ACP) are a group of nonspecific phosphohydrolases (Jaaska 1983). There may be a dozen or more of the phosphatase enzymes in plants (Gottlieb 1981). ACP is involved in the hydrolysis of phosphomonoesters; for example, in the formation of sucrose in photosynthesis (Goodman and Stuber 1983). It may also be involved in ripening and seed germination (Salinas and Benito 1984).

QUATERNARY STRUCTURE: Monomer (Nijenhuis 1971), dimer.
SUBCELLULAR LOCATION: Cytosol (Efron 1970).
REACTION: Through phosphate activity, naphthol is freed from α-naphthyl acid phosphate to be stained with Diazo blue/Fast Black K salt via the diazonium system.

Example 1

BUFFER: 0.05 M sodium acetate, pH 5.0 = 50 ml
α-NAP = 50 mg
Black K salt = 50 mg
INSTRUCTIONS: Incubate in the dark at 37°C; wash and fix.
SOURCE: Allen et al. (1963).

Example 2

BUFFER: Same as Example 1 above = 50 ml
Fast garnet GBC salt = 50 mg
$MgCl_2$ = 40 mg
α-NAP (sodium salt) = 50 mg
INSTRUCTIONS: Incubate in the dark at 30°C for 60 min; rinse and fix.
SOURCE: Scandalios (1969).

Example 3

BUFFER: 0.1 M sodium acetate-acetic acid, pH 5.0 = 50 ml
Diazo blue B = 150 mg
α-Naphthyl acid phosphate (NAP) = 150 mg
SOURCE: Cardy and Beversdorf (1984).
Note: β- or α-naphthyls may be used, but specificity of either one should be established for each electrophoretic system (Vallejos 1983). The naphthyls may be dissolved in a solvent (e.g. 50% acetone) before adding to the buffer.

10.4.3. Aconitase (ACO) E.C. 4.2.1.3 (Also called aconitate hydratase)

Aconitase (ACO) isozymes function in the Krebs cycle (TCA) in the interconversion of three tricarboxylic acids: citrate, *cis*-aconitase, and isocitrate (Zubay 1983). Koen and Goodman (1969) postulated four structural alleles at two loci to govern the genetics of the cytoplasmic ACO. Brown (1983) identified one locus with three alleles in the leaves of barley. The activity of the enzyme in plant extract is very sensitive to freezing and thawing (Stuber et al. 1988).

QUATERNARY STRUCTURE: Monomer (Doong and Kiang 1987).

SUBCELLULAR LOCATION: Cytoplasm (Dickman and Speyer 1954), mitochondria (Shepherd and Kalnitsky 1954).

REACTION:

$$\text{Aconitate} \xrightarrow{\text{ACO}} \text{isocitrate}$$

$$\text{Isocitrate} + \text{NADP}^+ \xrightarrow{\text{IDH}} \alpha\text{-Ketoglutarate} + \text{NADPH}$$

STAINING SYSTEM: The NADPH is used in the tetrazolium system.

Example 1

BUFFER: 1.0 M tris-HCl, pH 8.5	= 5 ml
Water	= 45 ml
cis-Aconitic acid (pH to 8.0)	= 35 mg
Isocitric dehydrogenase	= 3.5 units
1.0 M $MgCl_2$	= 0.5 ml
$NADP^+$	= 5 mg
MTT	= 2.5 mg
PMS	= 0.5 mg

SOURCE: A.L. Koen (1969, unpublished); Soltis et al. (1983).

Example 2

BUFFER: 0.2 M tris-HCl, pH 8.0	= 50 ml
cis-Aconitic acid (pH to 8.0)	= 100 mg
$MgCl_2$	= 50 mg
$NADP^+$	= 10 mg
MTT	= 10 mg
PMS	= 2 mg
Isocitrate dehydrogenase	= 20 mg

INSTRUCTIONS: Incubate in the dark at 30°C for 60 min. Rinse and fix.

SOURCE: Cardy and Beversdorf (1984).

Example 3

BUFFER: 0.2 M tris-HCl, pH 8.0	= 50 ml
Aconitic anhydride (pH to 8.0)	= 100 mg
$MgCl_2$	= 50 mg
$NADP^+$	= 10 mg
MTT	= 10 mg
PMS	= 2 mg
Isocitrate dehydrogenase	= 20 mg

(Aconitic anhydride is less expensive than the aconitic acid.)

SOURCE: Cardy and Beversdorf (1984); C. Sneller (personal communication).

10.4.4. Adenosine Deaminase (ADA) E.C. 3.5.4.4

BUFFER: 0.2 M tris-HCl, pH 8.0	= 1 ml
Water	= 4 ml
Xanthine oxidase	= 1.6 units
Nucleoside phosphorylase	= 5 units
Adenosine	= 40 mg
NBT	= 15 mg
PMS	= 50 mg
Sodium arsenate	= 50 mg

INSTRUCTIONS: Apply drops of staining solution to the gel surface with a Pasteur pipette; incubate at 37°C; rinse and fix.

SOURCE: Spencer et al. (1968).

10.4.5. Adenylate Kinase (ADK) E.C. 2.7.4.3

Adenylate kinase is one of a series of kinases (phosphotransferases) that convert mononucleotides to their metabolically active diphosphate and triphosphate form (Zubay 1983). ADK is especially active in tissues in which turnover energy from adenine nucleotides is high; for example, the mitochondria (Zubay 1983).

QUATERNARY STRUCTURE: Monomer (Stuber et al. 1988).

SUBCELLULAR LOCATION: Mitochondria (Zubay 1983), chloroplast.

REACTION:

$$\text{2ADP} \xrightarrow{\text{ADK}} \text{ATP} + \text{AMP}$$

$$\text{AMP} + \text{glucose} \xrightarrow{\text{Hexokinase}} \text{ADP} + \text{G-6-P}$$

$$\text{G-6-P} + \text{NADP}^+ \xrightarrow{\text{G-6-PD}} \text{6-PG} + \text{NADPH}$$

STAINING SYSTEM: NADPH is used in the tetrazolium staining system.

Example 1

BUFFER: 0.5 M tris-HCl, pH 7.1	= 5 ml
Water	= 45 ml
Glucose	= 45 mg
Adenosine diphosphate	= 10 mg
Hexokinase	= 80 units
Glucose-6-phosphate dehydrogenase	= 40 units
$MgCl_2.6H_2O$	= 10.5 mg
$NADP^+$	= 12.5 mg
PMS	= 1.5 mg
NBT	= 10 mg

INSTRUCTIONS: Incubate at 37°C; rinse and fix.

SOURCE: Fildes and Harris (1966); Shaw and Prasad (1970).

Example 2

Agar overlay

Solution A

BUFFER: 0.05 M tris-HCl, pH 8.0	= 15 ml

Hexokinase enzyme	= 78.12 units
NAD^+-dependent Glucose-6-phosphate dehydrogenase	= 37.5 units
$MgCl_2$	= 25 mg
NAD^+	= 5 mg
Adenosine diphosphate	= 20 mg
β-D(+)-Glucose	= 75 mg
MTT	= 5 mg
PMS	= 1.25 mg

Solution B

BUFFER: (same as above)	= 15 ml
Agar	= 2 mg

INSTRUCTIONS: Heat Solution B to dissolve agar, and then cool to 60°C before adding Solution A. Pour the complete staining solution on gel and incubate at 37°C for 60 min. Leave the gel overnight in a refrigerator.
SOURCE: Stuber et al. (1988).

10.4.6. Aldolase (ALD) E.C. 4.1.2.13

Aldolase catalyzes the reaction of the cleavage of fructose-1, 6-bisphosphate into two molecules of triose phosphate (Zubay 1983).
QUARTERNARY STRUCTURE: Tetramer.
SUBCELLULAR LOCATION: Chloroplast (Weeden and Gottlieb 1980).
REACTION:

$$\text{F-1,6-DP} \xrightarrow{\text{ALD}} \text{GA-3-P} + \text{DHAP}$$

$$\text{GA-3-P} + NAD^+ \xrightarrow{\text{GA-3-PD}} \text{1,3-DPG} + \text{NADH}$$

STAINING SYSTEM: The NADH is used in the tetrazolium staining system.

Example 1

BUFFER: 0.5 M tris-HCl, pH 7.1	= 5 ml
Water	= 45 ml
Fructose 1,6-diphosphate.$5H_2O$ (sodium salt)	= 272.5 mg
Glyceraldehyde-3-phosphate dehydrogenase	= 50 units
NAD+	= 25 mg
NBT	= 15 mg
PMS	= 1 mg
Arsenic acid (sodium salt)	= 75 mg

INSTRUCTIONS: Incubate at 37°C; rinse and fix.
SOURCE: Shaw and Prasad (1970).

10.4.7. Alkaline Phosphatase (AP) E.C. 3.1.3.1

The difference between AP and ACP is in the pH optima. AP is assayed under acid conditions whereas ACP is assayed under alkaline pH. (See also acid phosphatase.)
QUATERNARY STRUCTURE: Monomer.
SUBCELLULAR LOCATION: Cytosol.
REACTION: Same as for acid phosphatase.

Example 1

Water	= 50 ml
β-Naphthyl sodium phosphate	= 25 mg
$Mg_2SO_4.7H_2O$	= 61.5 mg
Fast blue RR	= 25 mg

INSTRUCTIONS: Incubate at 37°C; rinse and fix.
SOURCE: Boyer (1961).

Example 2

BUFFER: 50 mM tris, pH 8.5	= 50 ml
$MgCl_2$ 1 M	= 0.5 ml
$MnCl_2$ 1 M	= 0.5 ml
Na α-naphthyl phosphate 1% (dissolve in 50% acetone)	= 1.5 ml
Fast blue RR salt	= 50 mg

INSTRUCTIONS: Same as for example 1; Fast blue RR salt is very light sensitive.
SOURCE: Scandalios (1969).

10.4.8. Alcohol Dehydrogenase (ADH) E.C. 1.1.1.—

Alcohol dehydrogenase (ADH) enzymes are differentiated on the basis of their substrate specificities and type of cofactor required. NAD-dependent ADH (E.C. 1.1.1.1) acts on primary and secondary aliphatic alcohols whereas NAD-ADH (E.C. 1.1.1.90) acts on aromatic alcohols (Hart 1969). Jaaska (1978) reported a coenzyme-nonspecific ADH and an NADP-dependent ADH (E.C. 1.1.1.91) that act on aromatic alcohols.

ADH is most noted for its role in anaerobic respiration (alcohol fermentation), in which Johnson (1976) reported that it performs in a regulatory capacity. Anaerobic respiration is critical in periods of limited oxygen supply (e.g. seed dormancy and flooding). When samples for ADH assays are taken from plant materials under anaerobic conditions (flooded roots, seeds submerged in water for many hours), ADH activity tends to be enhanced.

ADH is encoded by two genes in most species. In wheat, Hart (1983) reported different types of ADH.
QUATERNARY STRUCTURE: Dimer (Stuber et al. 1988).
SUBCELLULAR LOCATION: Cytosol.
REACTION:

$$RCH_2OH + NAD^+ \xrightarrow{\text{ADK}} RCHO + NADH$$

(R = methyl group in the case of ethyl alcohol)
STAINING SYSTEM: NADH is used in the tetrazolium system.

Example 1

BUFFER: 0.1 M tris-HCl, pH 7.5	= 50 ml
Ethanol (add just before incubating)	= 3 ml
NAD^+	= 25 mg
MTT	= 10 mg
PMS	= 2 mg

INSTRUCTIONS: Incubate in the dark at 30°C for 15 to 60 min; rinse and fix.
SOURCE: Tanksley (1979).

Example 2

BUFFER: 0.05 M Sodium phosphate, pH 7.0	= 50 ml
Ethanol 95%	= 10 ml
NAD^+	= 60 mg
MTT	= 20 mg
PMS	= 2 mg

SOURCE: Original source not found.

Example 3

BUFFER: 0.05 M tris-HCl, pH 8.0	= 50 ml
95% Ethanol	= 1 ml
NAD^+	= 20 mg
MTT	= 20 mg
PMS	= 5 mg

SOURCE: Stuber et al. (1988).

10.4.9. Amylase (AMY) E.C. 3.2.1.—

Amylase is a starch-hydrolyzing enzyme. There are three types of amylases, the α- (E.C. 3.2.1.1) and β-amylases (E.C. 3.2.1.2) the more frequently investigated in research (Jaaska 1983). The third type is glycoamylase (E.C. 3.2.1.3). The different types of amylases are distinguished on the basis of their physical and chemical properties: α-amylase is intolerant of pH <3.6 and is activated by calcium, whereas the reverse is true for β-amylase (Frydenberg and Nielsen 1966).

QUARTERNARY STRUCTURE: Monomer (Brown 1983).

Example 1

Buffer 1:

50 mM Na Acetate, pH 5.6	= 50 ml
1 M $CaCl_2$	= 1 ml

Buffer 2:

10 mM I_2	= 25 ml
14 mM KI	= 25 ml

INSTRUCTIONS: Incubate slice of gel in staining Solution 1 (buffer 1) at 30–50°C for 1 hr, then discard solution and rinse gel with distilled water. Submerge gel in Solution 2; discard and score zymogram without delay.

SOURCE: Siepman and Stageman (1967).

Example 2

Overlay method: This method is used with polyacrylamide gel electrophoresis.

Medium:

Water	= 25 ml
Starch (soluble)	= 250 mg
Agar	= 375 mg

BUFFER:

0.1 M Na Acetate, pH 5.5	= 25 ml

Iodine solution:

10 mM I_2	= 25 ml
14 mM KI	= 25 ml

INSTRUCTIONS: Boil starch and agar suspension and cool to 45°C, and add acetate buffer. Mix and pour onto gel and incubate at 30°C for 1–2 hr. Separate overlay from the gel and score without delay.

SOURCE: Frydenberg and Nielsen (1966).

10.4.10. Arginine Aminopeptidase (AMP) E.C. 3.4.11.1

Example

STOCK BUFFER: pH 3.7
0.2 M tris
0.2 M Maleic acid

ASSAY BUFFER: pH 6.2; prepared by mixing 4 stock:$3H_2O$:2 (0.2 M NaOH)

STAIN BUFFER: AMP buffer pH 6.2	= 50 ml
L-Arginine-β-naphthylamide (dissolve in 5 ml *N,N′*-dimethyl formamide)	= 45 mg
$MgCl_2$	= 50 mg
Fast black K salt	= 25 mg

INSTRUCTIONS: Submerge the gel in the stain solution and incubate for 60 min. Remove from the incubator and leave at room temperature overnight. Rinse and score.

SOURCE: Stuber et al. (1988).

10.4.11. Aspartate Aminotransferase (AAT) E.C. 2.6.1.1 [Also called glutamate oxaloaceate transaminase (GOT)]

Aspartate aminotransferase (AAT) functions in amino group transport. Its role in transamination is to remove nitrogen from amino acids to produce keto acids for the Kreb's cycle and for gluconeogenesis (Goodman and Stuber 1983). AAT also serves as a link between carbohydrate and amino acid metabolism in the plant cell (Whight and Forrest 1978).

QUATERNARY STRUCTURE: Dimer (Arus and Orton 1983).

SUBCELLULAR LOCATION: Plastid, cytosol, mitochondria (Gorman et al. 1982).

REACTION:

$$\text{L-Aspartate} + \text{2-Oxoglutarate} \xrightarrow{\text{AAT}} \text{Oxaloacetate} + \text{L-Glutamate}$$

STAINING SYSTEM: Oxaloacetate enters the diazolium staining system where it reacts with fast blue BB salt (Babson et al. 1962).

Example 1

BUFFER: 0.1 M Potassium phosphate, pH 7.0	= 50 ml
L-Aspartic acid	= 266 mg
α-Ketoglutaric acid	= 36.5 mg
Pyridoxal-5-phosphate	= 25 mg
Fast violet B salt	= 100 mg

INSTRUCTIONS: Incubate at 37°C; fix in glycerin.

SOURCE: Schwartz et al. (1964); Shaw and Prasad (1970).

Example 2

BUFFER: 0.1 M tris-HCl, pH 8.3	= 50 ml
L-Aspartic acid	= 50 mg
α-ketoglutaric acid	= 50 mg
Fast blue BB	= 75 mg
Pyridoxal-5-phosphate	= 2.5 mg

INSTRUCTION: Stain in the dark at 30°C; rinse and fix.

SOURCE: Selander et al. (1971).

Example 3

Substrate solution, pH 7.4	= 25 ml
Water	= 25 ml
Fast blue BB	= 50 mg
Substrate solution:	
α-ketoglutaric acid	= 146.1 mg
L-aspartic acid	= 532.4 mg
EDTA	= 200 mg
Sodium phosphate (monobasic)	= 5.7 g
PVP-40	= 2.0 g
Water	= 200.0 ml

INSTRUCTIONS: Incubate at 30°C for 45 min; rinse and fix.
SOURCE: Cardy and Beversdorf (1984).

10.4.12. Catalase (CAT) E.C. 1.11.1.6

Catalase is a redox enzyme that catalyzes dismutation of hydrogen peroxide.
QUATERNARY STRUCTURE: Tetramer (Scandalios 1974; Stuber et al. 1988).
SUBCELLULAR LOCATION: Mitochondria, cytosol (Goodman and Stuber 1983).
REACTION:

$$2H_2O_2 \xrightarrow{\text{CAT}} 2H_2O + O_2$$

$$H_2O_2 + I^- \longrightarrow I_2$$

Iodide is oxidized to iodine, resulting in a blue precipitate.

Example 1

Solution A:	
Na thiosufate 60 mM	= 15 ml
Hydrogen peroxide 3%	= 35 ml
Solution B:	
Potassium iodide 90 mM	= 50 ml
Glacial acetic acid	= 0.25 ml

INSTRUCTIONS: Mix Solution A very quickly just before pouring it onto the gel; incubate for 30 sec and drain Solution A; add Solution B. This is a negative stain, hence CAT activity results in achromatic spots on a dark background. The spots may fade quickly in some cases, so score without delay.
SOURCE: Thorup et al. (1961).

Example 2

Gel wash:
BUFFER:

0.3 M Boric acid, pH = 6.5 (pH with 1 M HCl)	= 100 ml

Soak the gel slice at 0°C for 45 min.
Stain:

A.	Potassium iodide	= 2 g
	Acetic acid	= 2 ml
	Water	= 100 ml
B.	Water	= 100 ml
	3% Hydrogen peroxide	= 1 ml

INSTRUCTIONS: Soak the gel in Solution A for 60 sec and wash 3 times; add Solution B. Incubate until bands (white) appear. Fix in 50% glycerin.
SOURCE: Robinson (1966).

Example 3

Water	= 50 ml
Potassium iodide	= 100 mg
Glacial acetic acid	= 0.5 ml

INSTRUCTIONS: Refrigerate the gel slice for 15 min; flood with 0.6% hydrogen peroxide for 60 sec and drain and rinse with water. Add the stain and swirl gently until bands (achromatic) appear.
SOURCE: Cardy and Beversdorf (1984).

Example 4

Water	= 50 ml
Potassium ferricyanide	= 500 mg
Ferric chloride	= 500 mg

INSTRUCTIONS: Pour 0.01% hydrogen peroxide on the gel and swirl; let mixture sit for 5 min. Rinse with distilled water and add the stain. Rinse after 5 min and store in water.
SOURCE: Stuber et al. (1988).

10.4.13. Creatine Kinase (CK) E.C. 2.7.3.2

REACTION:
$MgATP^{2} + creatine^{+} \Leftrightarrow MgADP^{-} + phosphocreatine^{-} + H^{+}$

BUFFER: 0.5 M tris-HCl, pH 7.1	= 10 ml
Water	= 90 ml
Creatine phosphate	= 731 mg
ADP	= 75 mg
Glucose	= 21 mg
$NADP^+$	= 25 mg
Hexokinase	= 160 units
Glucose-6-phosphate dehydrogenase	= 80 units
PMS	= 3 mg
NBT	= 20 mg

INSTRUCTIONS: Incubate at 37°C until dark blue bands appear. Rinse and fix.
SOURCE: Dawson et al. (1966); Shaw and Prasad (1970).

10.4.14. Diaphorase (DIA) E.C. 1.6.—.—

Diaphorases are a widespread group of enzymes that catalyze the oxidation of β-NADH or β-NADPH in the presence of an electron acceptor. They are assayed by using artificial dyes (*Worthington Manual* 1968). The E.C. number for diaphorase varies. In soybean, Gorman and Kiang (1977) reported monomeric and tetrameric forms with E.C. 1.6.4.3. In spinach, Dixon and Webb (1979) found monomers and dimers of the enzyme with E.C. 1.6.99.1, and monomers have been reported in barley with E.C. 1.6.99.3 (Brown 1983).
QUATERNARY STRUCTURE: Monomer (Gorman and Kiang 1977); tetramer (Dixon and Webb 1979).
SUBCELLULAR LOCATION: Mitochondria, cytosol (Gorman and Kiang 1977).
REACTION:

$$\text{Oxidized DCPIP} + \text{NADH} \xrightarrow{\text{DIA}} \text{Reduced DCPIP} + \text{NAD}$$

Example 1
Agar overlay
Solution A:
Water = 25 ml
Agar = 375 mg
Solution B:
0.1 M tris, pH 8.0 = 25 ml
2,6-dichlorophenol indolphenol (DCPIP) = 2 mg
NADH = 6 mg
INSTRUCTION: Incubate gel with overlay at 30°C for 15 to 30 min. Bands diffuse and fade quickly so photograph or score without delay.
SOURCE: Brewer et al. (1967).

Example 2
BUFFER: 0.2 M tris-HCl, pH 8.0 = 50 ml
Menadione = 20 mg
NADH = 10 mg
MTT = 10 mg
SOURCE: Original source not found.

Example 3
BUFFER: 0.2 M tris-HCl, pH 8.0 = 50 ml
DCPIP = 3.0 mg
NADH = 10 mg
MTT = 10 mg
INSTRUCTIONS: Incubate in the dark for 30 min. Rinse and fix.
SOURCE: Cardy and Beversdorf (1984).

Example 4
Agar overlay
Solution 1:
BUFFER: 0.05 M tris-HCl, pH 8.0 = 15 ml
DCPIP = 1 mg
NADH = 10 mg
MTT = 5 mg
Solution 2:
BUFFER: Same as for solution 1. = 15 ml
Agar = 200 mg
INSTRUCTIONS: Dissolve agar by heating and cool to 60°C. Add solution 1 and mix. Pour the complete solution over the gel and incubate for 2 hr or more. Keep overnight in refrigerator.
SOURCE: Stuber et al. (1988).

10.4.15. Endopeptidase (ENP) E.C. 3.4.23.6

Endopeptidases are a group of enzymes that characteristically degrade proteins by acting on peptidases from within. They are also associated with functions such as seed germination, aging, and fruit ripening.
QUATERNARY STRUCTURE: Monomer (Stuber et al. 1988).

Example
BUFFER: 0.1 M tris-Maleic acid, pH 5.5 = 50 ml
BANA (β-*N*-benzoyl-DL-arginine-α-naphthylamide HCl = 5 mg
Black K salt = 5 mg
$MgCl_2$ = 25 mg
INSTRUCTIONS: Incubate the gel for 60 min. Rinse and fix.
SOURCE: Cardy and Beversdorf (1984).

10.4.16. Enolase (ENO) E.C. 4.2.1.11

Example
BUFFER: 0.2 M tris-HCl, pH 8.0 = 1 ml
Water = 4 ml
Pyruvate kinase = 20 units
Lactate dehydrogenase = 20 mg
2-Phospho-D-glycerate (Na salt) = 25 mg
NADH = 15 mg
ADP = 2.5 mg
$MgCl_2$ = 20 mg
Fructose-1,6-diphosphate (tetrasodium salt) = 1 mg
INSTRUCTIONS: Apply drops of staining solution to the gel surface using a Pasteur pipette. View the gel under longwave u.v. light. Enolase appears as defluoresced bands against a fluorescent background. Score immediately.
SOURCE: Omenn and Cohen (1971).

10.4.17. Esterase (EST) E.C. 3.1.1.—

The esterases are a group of enzymes highly nonspecific in their action in plants, thus making distinction among types a difficult task (Goodman and Stuber 1983). The particular carboxylic ester hydrolases that are stained on a gel depends on the composition of the staining solution. McCleod and Brewbaker (1975) reported ten loci that have been genetically verified in maize, some of which migrate anodally whereas others migrate cathodally when electrophoresis is conducted with a buffer pH of between 8.2 and 8.6.
QUARTERNARY STRUCTURE: Monomer (Tanksley and Rick 1980); dimer (Stuber et al. 1988).
REACTION: This is a nonspecific system in which the substrates may have little to do with the biological or natural substrate. In staining, α-naphthol is coupled with a diazonium salt (blue RR) after the α-naphthol is liberated from α-naphthyl acetate by esterase activity.

Example 1
Colorimetric esterase
BUFFER: 1.0 M Phosphate, pH 6.0 = 5 ml
Water = 45 ml
α-Naphthyl acetate = 20 mg
β-Naphthyl acetate (dissolve in acetone) = 20 mg
Fast blue RR salt = 50 mg
INSTRUCTIONS: Incubate at room temperature.
SOURCE: Gottlieb (1974); Soltis et al. (1983).

Example 2
BUFFER: 0.05 M Potassium phosphate (monohydrate) pH = 6.0 = 50 ml
N-propanol = 2.5 ml
β-Naphthyl acetate = 20 mg
Fast garnet GBC salt = 25 mg
INSTRUCTIONS: Pour staining solution on the gel. After five min add 30 mg α-Naphthyl acetate and incubate for 45 min. Rinse and fix.
SOURCE: Stuber et al. (1988).

Example 3
Substrate
1% α- and β-Naphthyl acetate
α-Naphthyl acetate = 1 g
β-Naphthyl acetate = 1 g
Acetone = 50 ml
Water = 50 ml
Stain
BUFFER: 1% 0.5 M tris-HCl, pH = 7.1 = 5 ml
Water = 43.5 ml
1% α- and β-Naphthyl acetate = 1.5 ml
Fast blue RR salt = 50 mg
INSTRUCTIONS: Incubate at room temperature. Wash and fix.
SOURCE: Shaw and Prasad (1970).

Example 3
Fluorescent esterase
BUFFER: 1.0 M Sodium acetate, pH 5.0 = 9 ml
Water = 26.5 ml
4-methylumbelliferyl acetate = 2 mg
(dissolve in 12.5 ml acetone)
INSTRUCTIONS: Stain in the dark at room temp. View under longwave u.v. light.
SOURCE: Mitton et al. (1979).

10.4.18. Fructose-1,6-diphosphatase (F6DP) E.C. 3.1.3.11

QUATERNARY STRUCTURE: Tetramer.
REACTION:

$$\text{Fructose-diphosphate} \xrightarrow{\text{FDP}} \text{F-6-P}$$

$$\text{F-6-P} \xrightarrow{\text{GPI}} \text{G-6-P}$$

$$\text{F-6-P} + \text{NADP}^+ \xrightarrow{\text{G-6-PD}} \text{6PGA} + \text{NADPH}$$

STAINING SYSTEM: The NADPH is used in the tetrazolium staining system.

Example 1
BUFFER: 0.5 M tris-HCl, pH 7.5 = 5 ml
Water = 45 ml
$NADP^+$ = 7.5 mg
$Mg_2SO_4.7H_2O$ = 125 mg
PMS = 1 mg
NBT = 10 mg
2-Mercaptoethanol = .001 ml
Fructose 1,6-phosphate = 10 mg
Just before use, add the following:
Phosphohexose isomerase = 20 units
Glucose-6-phosphate dehydrogenase = 20 units
INSTRUCTIONS: Incubate the gel at 37°C. Rinse and fix.
SOURCE: D. Stout (1969, unpublished).

Example 2
BUFFER: 1.0 M tris-HCl, pH 8.0 = 5 ml
Water = 45 ml
Fructose-1,6-phosphate (trisodium salt) = 50 mg
1.0 M $MgCl_2$ = 0.5 ml
Phosphoglucoisomerase = 25 units
Glucose-6-phosphate dehydrogenase = 25 units
$NAD(P)^+$ = 5 mg
MTT = 2.5 mg
PMS = 0.5 mg
INSTRUCTIONS: Incubate in the dark at room temp. Rinse and fix.
SOURCE: Modification of Soltis et al. (1983).

10.4.19. Fumarase (FUM) E.C. 4.2.1.2 (Also called fumarase hydratase)

Fumarase catalyzes the reaction in which fumarate is converted to L-malate (Zubay 1983).
QUATERNARY STRUCTURE: Tetramer
REACTION:

$$\text{Fumarate} + H_2O \xrightarrow{\text{FUM}} \text{Malate}$$

$$\text{Malate} + \text{NAD}^+ \xrightarrow{\text{MDH}} \text{Oxaloacetate} + \text{NADH}$$

STAINING SYSTEM: NADH is used in the tetrazolium staining system.

Example 1
BUFFER: Sodium phosphate 50 mM, pH 7.0 = 50 ml
Fumaric acid (disodium salt) = 0.8 g
NAD^+ = 15 mg
MTT = 10 mg
PMS = 2 mg
Malate dehydrogenase = 20 units
INSTRUCTIONS: Incubate in the dark at 30°C for 1–2 hr. Rinse and fix.
SOURCE: Brewer and Sing (1970).

Example 2
BUFFER: 0.1 M Phosphate, pH 7.1 = 10 ml
Water = 40 ml
K Fumarate = 385 mg
NAD^+ = 40 mg
NBT = 15 mg
PMS = 0.5 mg
Malate dehydrogenase = 100 units
INSTRUCTIONS: Incubate the gel at 37°C. Rinse and fix.
SOURCE: Shaw and Prasad (1970).

10.4.20. β-D-Galactosidase (β-GAL) E.C. 3.2.1.23

β-D-Galactosidase (β-D-GAL) is believed to be involved in plant growth and differentiation and germination processes.
REACTION: β-D-GAL catalyzes the hydrolysis of terminal nonreducing β-D-galactose units in β-galactosides.

Example
BUFFER: 0.1 M Sodium phosphate pH 7.0 = 50 ml
4-Methylumberlliferyl-β-D-galactoside = 100 mg
INSTRUCTIONS: For a filter paper overlay, soak a strip of filter in the staining solution and lay it on the gel. Incubate for 30 min, remove the paper and score promptly under u.v. light.
SOURCE: Hughes (1981). (To stain for α-D-Galacto-

sidase, substitute the β form of the substrate with the α form).

10.4.21. Glucose-6-phosphate Dehydrogenase (G6PDH) E.C. 1.1.1.49

Glucose-6-phosphate dehydrogenase is involved in sugar metabolism and is the first enzyme in the phosphogluconate pentose phosphate pathway (Lehninger 1982).

QUATERNARY STRUCTURE: Dimer (Schnarrenberger and Oesser 1974).

SUBCELLULAR LOCATION: Cytosol, plastid (Schnarrenberger and Oesser 1974).

REACTION:

$$\text{G-6-P} + \text{NADP}^+ \xrightarrow{\text{G-6-PD}} \text{6-PG} + \text{NADPH}$$

STAINING SYSTEM: NADPH is used in the tetrazolium staining system.

Example 1

BUFFER: 0.1 M tris, pH 7.5	= 50 ml
1 M $MgCl_2.6H_2O$	= 0.5 ml
Glucose-6-phosphate	= 20 mg
$NADP^+$	= 7.5 mg
NBT	= 10 mg
PMS	= 2 mg

INSTRUCTIONS: Incubate in the dark at 30°C for 15–60 min. Rinse and fix.

SOURCE: Sing and Brewer (1969).

Example 2

BUFFER: 1.0 M tris-HCl, pH 8.0 or 8.5	= 5 ml
Water	= 45 ml
Glucose-6-phosphate (disodium salt)	= 50 mg
$NADP^+$	= 10 mg
MTT (or NBT)	= 5 mg
PMS	= 1 mg

INSTRUCTIONS: Stain in the dark at 37°C. Rinse and fix.

SOURCE: Shaw and Prasad (1970); Soltis et al. (1983).

10.4.22. Glucose-1-phosphate Transferase (G1PT)

Example

BUFFER: 0.1 M tris-Acetate, pH = 6.5	= 50 ml
0.1 M $MgCl_2$	= 2 ml
Glucose-1-phosphate	= 20 mg
α-D-Glucose	= 40 mg
Glucose-6-phosphate dehydrogenase	= 20 units
$NADP^+$	= 8 mg
MTT	= 8 mg
Meldola's blue	= trace

SOURCE: Original source not found.

10.4.23. β-D-Glucosidase (β-GLU) E.C. 3.2.1.21

β-D-Glucosidase is involved in some host-pathogen relationships in maize. It may also play a role in carbohydrate metabolism in plants. The *Glu1* may be the most variable locus known in plants (more than 30 alleles in maize) (Goodman and Stuber 1983).

QUATERNARY STRUCTURE: Dimer (Stuber et al. 1988).

SUBCELLULAR LOCATION: Cytoplasm (Goodman and Stuber 1983).

REACTION: The enzyme is stained through the diazonium system after β-D-glucose is hydrolyzed from the nonreducing termini of β-D-glucans.

Example 1

BUFFER: 0.05 M Sodium phosphate, pH 6.5	= 50 ml
6-Bromo-2-Naphthyl-β-D-glucose 1% (dissolve in acetone)	= 2.5 ml
Fast blue BB	= 50 mg

INSTRUCTIONS: Incubate in the dark at 30°C for 4–8 hr. Rinse and fix. Resolution is significantly improved when histidine-citrate gel buffers at low pH (5.7) are used.

SOURCE: Stuber et al. (1988).

Example 2

Solution A:

BUFFER: 0.05 M Potassium phosphate, pH 6.5	= 50 ml
PVP-40	= 1 g
Fast blue BB salt (just before staining)	= 100 mg

Solution B:

6-Bromo-2-Naphthyl-β-D-Glucoside	= 50 mg
N,N'-Dimethyl formamide (solvent)	= 5 ml

INSTRUCTIONS: Add Solutions A to B and pour on the gel. Incubate the gel for 60 min and let it stay outside at room temperature overnight. Rinse and fix.

SOURCE: Stuber et al. (1988).

10.4.24. Fluorescent β-D-Glucosidase (β-GLU) E.C. 3.2.1.21

This enzyme is the same as β-GLU above except that the stain is less permanent and the results appear much faster. The enzymes observed are the same (Stuber et al. 1988).

Example

Agar overlay

Solution A:

BUFFER: 0.05 M Potassium phosphate, pH 6.5	= 15 ml
4-Methylumbeliferyl-β-glucoside	= 7.5 mg

Solution B:

BUFFER: Same as for Solution A.	
Agar	= 200 mg

INSTRUCTIONS: Dissolve the agar by heating, and then cool to 60°C. Add Solution A to the agar, mix and pour over the gel. Score immediately using longwave u.v. light.

SOURCE: Stuber et al. (1988).

10.4.25. Glyceraldehyde-3-phosphate Dehydrogenase (G3PDH) E.C. 1.2.1.12

Example

Substrate solution

BUFFER: 0.2 M tris-HCl, pH 8.0	= 2 ml

Water = 3 ml
Aldolase = 100 units
Fructose-1,6-diphosphate (tetrasodium salt) = 270 mg
Incubate solution at 37°C for 30–60 min.
Staining solution
BUFFER: 0.2 M tris-HCl, pH 8.0 = 5 ml
Water = 35 ml
Substrate solution = 5 ml
NAD^+ = 25 mg
NBT = 15 mg
PMS = 1 mg
Sodium arsenate = 75 mg
INSTRUCTIONS: Incubate the gel in the staining solution at 37°C in dark. Rinse and fix.
SOURCE: Wright et al. (1972).

10.4.26. Glutamate dehydrogenase (GDH) E.C. 1.4.1.2

Glutamate dehydrogenase, a ubiquitous and substrate-specific enzyme, was reported as a hexamer in rice (Endo and Morishma 1983). On the basis of the cofactor requirement, two forms of GDH occur in higher plants, one NAD-dependent and one NADP-dependent.
QUATERNARY STRUCTURE: Monomer (McLeod et al. 1983); hexamer (Quiros 1983).
SUBCELLULAR LOCATION: Chloroplast, mitochondria.
REACTION:

$$\text{Glutamate} + NAD^+ + H_2O \xrightarrow{\text{GDH}} \alpha\text{-oxoglutarate} + NADH + NH_4^+$$

Example 1
BUFFER: 0.1 M tris, pH 7.5 = 50 ml
10 mM $CaCl_2$ = 0.1 ml
Sodium glutamate = 400 mg
NAD^+ = 15 mg
NBT = 10 mg
PMS = 2 mg
INSTRUCTIONS: Incubate in the dark at 30°C. Rinse and fix. Glutamic acid with the pH adjusted may be used.
SOURCE: Hartman et al. (1973).

Example 2
BUFFER: 1.0 M tris-HCl, pH 8.0 = 5 ml
Water = 35 ml
1.0 M L-Glutamic acid (free acid or monosodium), pH 8.0 = 10 ml
NAD^+ = 10 mg
MTT (or NBT) = 5 mg
PMS = 1 mg
INSTRUCTIONS: Stain in the dark at room temperature. Rinse and fix.
SOURCE: Gottlieb (1973); Soltis et al. (1983).

10.4.27. Glutamate-pyruvate-transaminase (GPT) E.C. 2.6.1.2

Example
BUFFER: 0.2 M tris-HCl, pH 8.0 = 1 ml
Water = 4 ml
Lactate dehydrogenase = 150 units
NADH = 15 mg
l-Alanine = 20 mg
α-Ketoglutaric acid = 10 mg
INSTRUCTIONS: Apply drops of staining solution to the gel surface with Pasteur pipette. View under long-wave u.v. GPT bands appear as defluoresced bands over a fluoresced background.
SOURCE: Chen et al. (1972).

10.4.28. β-Glucuronidase (β-GUN) E.C. 3.2.1.31

REACTION:

$$\beta\text{-D-Glucuronide} + H_2O \longrightarrow \text{An alcohol} + \text{D-glucuronate (fluorescent)}$$

Example 1
Substrate
BUFFER: 0.1 M Phosphate-citrate, pH 4.95 = 20 ml
6-Bromo 2-naphthyl β-D-glucuronide = 30 mg
Absolute ethanol = 10 ml
Water = 70 ml
Stain
BUFFER: 0.02 M Phosphate, pH 7.5 = 100 ml
Fast blue B = 100 mg
INSTRUCTIONS: Filter the staining solution before using and use fresh. Incubate in substrate buffer at 37°C overnight. Rinse with water and add the staining solution. Wait until blue bands appear. Wash with cold, distilled water (two times) and rinse with 0.1% acid solution.
SOURCE: Fondo and Bartalos (1969).

Example 2
Filter paper overlay
Staining solution:
Na acetate 50 mM, pH 4.5 = 100 ml
4-Methylumbellyferyl glucuronide 1% (in DMSO) = 5 ml
INSTRUCTIONS: Saturate the filter paper with the staining solution and lay it on top of gel. Incubate for 30 min, remove the paper, and view the gel under longwave u.v. light. Score immediately.
SOURCE: Hughes (1981).

10.4.29. Gluthione reductase (GR) E.C. 1.6.4.2

Gluthione is involved in lipid biosynthesis.
QUATERNARY STRUCTURE: Dimer.
REACTION:

$$GSSG + NADPH \longrightarrow NAPD^+ + 2GSH$$

Example
Agar overlay
Solution A:
BUFFER: 0.2 M tris, pH 8.0 = 50 ml

Na_2EDTA	= 1.25 g
Dithiobis	= 35 mg
GSSG	= 350 mg
NADPH	= 35 mg

Solution B:

H_2O	= 50 ml
Agar	= 750 mg

INSTRUCTIONS: Heat the mixture of EDTA and dithiobis in the buffer to dissolve. Cool to 45°C and add GSSG and NADPH. Boil the agar suspension to dissolve and cool to 45°C. Mix the agar with Solution A and pour onto the gel. Incubate for 1–2 hr in the dark and watch for yellow bands.
SOURCE: Brewer and Sing (1970).

10.4.30. Glycerate-2-dehydrogenase (G2D) E.C. 1.1.1.29

Example

BUFFER: 0.2 M tris-HCl, pH 8.0	= 50 ml
DL-Glyceric acid	= 100 mg
NAD^+	= 15 mg
MTT	= 15 mg
PMS	= 1 mg

SOURCE: O'Malley et al. (1980).

10.4.31. Hexokinase (HEX) E.C. 2.7.1.1 (Also called glucokinase)

Hexokinase catalyzes the phosphorylation reaction that brings free glucose (e.g. from the hydrolysis of sucrose, starch) into the hexose phosphate pool (Zubay 1983).
QUATERNARY STRUCTURE: Monomer (Stuber et al. 1988).
SUBCELLULAR LOCATION: Cytosol, plastid, mitochondria.
REACTION:

$$\text{Glucose} + \text{ATP} \xrightarrow{\text{HEX}} \text{G-6-P} + \text{ADP}$$

$$\text{G-6-P} + \text{NADP}^+ \xrightarrow{\text{G-6-PD}} \text{6-PG} + \text{NADPH}$$

STAINING SYSTEM: NADPH is used in the tetrazolium system.

Example 1

BUFFER: 0.5 M tris-HCl, pH 7.1	= 5 ml
Water	= 45 ml
Glucose	= 45 mg
$MgCl_2.6H_2O$	= 10.5 mg
ATP	= 12.5 mg
$NADP^+$	= 12.5 mg
PMS	= 1.5 mg
NBT	= 10 mg
Glucose-6-phosphate dehydrogenase	= 40 units

INSTRUCTIONS: Incubate the gel at 37°C. Rinse and fix.
SOURCE: Eaton et al. (1966).

Example 2

BUFFER: 1.0 M tris-HCl, pH 8.5	= 5 ml
Water	= 45 ml
Glucose	= 45 mg
1.0 M $MgCl_2$	= 2.5 ml
EDTA (tetrasodium salt, dihydrate)	= 20 mg
$NAD(P)^+$	= 5 mg
MTT (or NBT)	= 7.5 mg
Glucose-6-phosphate dehydrogenase	= 20 units
PMS	= 1 mg
ATP	= 12.5 mg

INSTRUCTIONS: Stain in the dark at room temperature.
SOURCE: Modification by Soltis et al. (1983).

Example 3
Agar overlay
Solution 1:

BUFFER: 0.05 M tris-HCl, pH 8.0	= 15 ml
β-D(+)-Glucose	= 125 mg
Adenosine-5-triphosphate	= 125 mg
$MgCl_2$	= 50 mg
NAD^+	= 10 mg
Glucose-6-phosphate dehydrogenase (NAD dependent)	= 56.25 units
MTT	= 5 mg
PMS	= 1.25 mg

Solution 2:

BUFFER: Same as for Solution 1	= 15 ml
Agar	= 200 mg

INSTRUCTIONS: Heat Solution 2 until agar dissolves, then cool to 60°C. Add Solution 1 and swirl. Apply over the gel and incubate for 2 hr. Refrigerate overnight; store in refrigerator.
SOURCE: Stuber et al. (1980).

10.4.32. Isocitrate dehydrogenase ($NADP^+$) (IDH) E.C. 1.1.1.42

IDH has a role in the TCA cycle. It is involved in the movement of reducing power between mitochondria and the cytosol (Lehninger 1982) and catalyzes the oxidation of isocitrate to α-ketoglutarate.
QUATERNARY STRUCTURE: Dimer (Goodman and Stuber 1988).
SUBCELLULAR LOCATION: Cytosol (Ni et al. 1987).
REACTION:

$$\text{Isocitrate} + \text{NADP}^+ \xrightarrow{\text{IDH}} \text{Oxalosuccinate} + \text{NADPH}$$

STAINING SYSTEM: The NADPH is used in the tetrazolium staining system.

Example 1

BUFFER: 0.1 M tris, pH 7.5	= 50 ml
1 M $MnCl_2$	= 0.5 ml
DL-Isocitrate (trisodium salt)	= 50 mg
$NADP^+$	= 7.5 mg
MTT	= 10 mg
PMS	= 2 mg

INSTRUCTIONS: Incubate in the dark at 30°C. Rinse and fix.
SOURCE: Fine and Costello (1963).

Example 2

BUFFER: 1.0 M tris-HCl, pH 8.0	= 5 ml

Water	= 42.5 ml
Isocitric acid (trisodium salt)	= 50 mg
1.0 M $MgCl_2$	= 2.5 ml
$NADP^+$	= 5 mg
MTT (or NBT)	= 7.5 mg
PMS	= 1 mg

INSTRUCTIONS: Stain in the dark at room temperature.
SOURCE: Henderson (1965); Soltis et al. (1983).

Example 3
Agar overlay
Solution 1:

BUFFER: 0.05 M tris-HCl, pH 8.0	= 15 ml
$MgCl_2$	= 50 mg
DL-Isocitric acid	= 150 mg
$NADP^+$	= 5 mg
NBT	= 5 mg
PMS	= 1 mg

Solution 2:

BUFFER: Same as in Solution 1	= 15 ml
Agar	= 200 mg

INSTRUCTIONS: Heat Solution 2 to dissolve the agar and cool to 60°C. Add Solution 1 and swirl; apply solution to the gel. Incubate for 60 min. Stain develops overnight. Rinse and fix.
SOURCE: Stuber et al. (1988).

10.4.33. Lactate Dehydrogenase (LDH) E.C. 1.1.1.27

Lactate dehydrogenase is active in animal tissue. Honold et al. (1966) reported it to be also active in wheat seeds.
SUBCELLULAR LOCATION: Cytosol (Endo and Morishma 1983).
REACTION:

L-lactate + NAD^+ ⟶ pyruvate + NADH

STAINING SYSTEM: NADH is used in the tetrazolium staining system.

Example 1

BUFFER: 0.5 M tris-HCl, pH 7.1	= 15 ml
NAD^+	= 50 mg
NBT	= 30 mg
PMS	= 2 mg
H_2O = up	to 100 ml
0.1 M NaCN	= 5 ml
Substrate: (1 M Na DL-lactate pH 7.0)	= 10 ml
85% DL-lactic acid	= 10.6 ml
1 M $Na_2CO_3.H_2O$	= 49 ml

INSTRUCTIONS: Keep the flask cool while mixing solution.
SOURCE: Shaw and Prasad (1970).

Example 2

BUFFER: 0.1 M tris, pH 7.5	= 50 ml
1.0 M $MgCl_2.6H_2O$	= 1 ml
DL-lactate 85%	= 10 ml
NAD^+	= 30 mg
PMS	= 4 mg

INSTRUCTIONS: Add the DL-lactate just before incubation. Incubate in the dark at 30°C for 30–60 min.
SOURCE: Shaw and Prasad (1970); Vallejos (1983).

10.4.34. Leucine Aminopeptidase (LAP) E.C. 3.4.11.—

The precise name of this enzyme is aminopeptidase. I.U.B (1979) reports a number of aminopeptidases: E.C. 3.4.11.1 is found in the cytosol and is activated by heavy metals whereas E.C. 3.4.11.2 is microsomal in location and is not activated by heavy metals.
QUATERNARY STRUCTURE: Monomer.
SUBCELLULAR LOCATION: Cytosol.
REACTION:

Peptide ⟶ L-amino acid

L-amino acid $\xrightarrow{\alpha\text{-amino acid oxidase}}$ H_2O_2

H_2O_2 $\xrightarrow{\text{peroxidase}}$ *o*-dianisidine

Example 1
BUFFER: tris-Maleate, pH 6.0
Solution A:

Tris	= 24.2 g
Maleic acid	= 23.2 g

Solution B:

0.2 M NaOH	= 26 ml

Add: 50 ml A + 26 ml B and make up to 200 ml with water
Stain solution:

Buffer	= 25 ml
Water	= 25 ml
Black K salt	= 25 mg
L-Leucyl β-naphthylamide	= 10 mg

INSTRUCTIONS: Incubate the gel at 37°C. Wash and fix.
SOURCE: Smith and Rutenberg (1966); Shaw and Prasad (1970).

Example 2

BUFFER: 1.0 M Phosphate, pH 6.0	= 5 ml
Water	= 45 ml
Black K salt (or fast black salt)	= 25 mg
L-Leucine-β-naphthylamide (free base or acid salt); dissolve in *N,N′* dimethylformamide	= 10 mg

INSTRUCTIONS: Stain in the dark at room temperature. Rinse and fix.
SOURCE: Gottlieb (1973); Soltis et al. (1983).

10.4.35. Malate Dehydrogenase (MDH) E.C. 1.1.1.37

Malate dehydrogenase catalyzes the conversion of malate to oxaloacetate. The enzyme exhibits subcellular compartmentalization, each with unique isozymes (Gottlieb 1981). Tyson et al. (1986) reported that MDH is subject to post-translational modification, an event that complicates zymogram analysis.
QUATERNARY STRUCTURE: Dimer (Longo and Scandalios 1969).
SUBCELLULAR LOCATION: Microbes, mitochondria, cytosol

(Ting et al. 1975).

REACTION:

$$\text{Malate} + \text{NAD}^+ \xrightarrow{\text{MDH}} \text{Oxaloacetate} + \text{NADH}$$

STAINING SYSTEM: The NADH is used in the tetrazolium staining system.

Example 1

BUFFER: 0.1 M tris, pH 7.5	= 50 ml
DL-Malate 1 M, pH 7.5	= 1.5 ml
NAD^+	= 15 mg
MTT	= 10 mg
PMS	= 2 mg

INSTRUCTIONS: Incubate the gel in the dark at 30°C. Rinse and fix.

SOURCE: Brown et al. (1978).

Example 2

Substrate:

BUFFER: Na L-malate, pH 7.0	
L-Malic acid	= 13.4 g
2 M $Na_2CO_3.H_2O$	= 49 ml
Water	= 1 l

Keep the flask cool while mixing solution.

Stain:

BUFFER: 0.5 M tris-HCl, pH 7.1	= 7.5 ml
Water	= 35 ml
1 M Na L-malate pH 7.0	= 5 ml
0.1 M NaCN	= 2.5 ml
NAD^+	= 25 mg
NBT	= 15 mg
PMS	= 1 mg

SOURCE: Shaw and Koen (1967).

Example 3

BUFFER: 1.0 M tris-HCl, pH 8.0 of 8.5	= 5 ml
2.0 M DL-Malic acid (pH with NaOH to 8.0)	= 5 ml
Water	= 40 ml
NAD^+	= 5 mg
MTT or NBT	= 5 mg
PMS	= 1 mg

INSTRUCTIONS: Stain in the dark at room temperature. Rinse and fix.

SOURCE: Shaw and Prasad (1970); Soltis et al. (1983).

10.4.36. Malic enzyme (ME) E.C. 1.1.1.40

Malic enzyme is known to undergo changes in expression in development.

QUATERNARY STRUCTURE: Tetramer (Aicher 1988).

SUBCELLULAR LOCATION: Cytosol.

REACTION:

$$\text{Malate} + \text{NADP}^+ \xrightarrow{\text{ME}} \text{Pyruvate} + \text{NADPH}$$

STAINING SYSTEM: NADPH is used in the tetrazolium staining system.

Example 1

BUFFER: 1.0 M tris-HCl, pH 8.0 or 8.5	= 5 ml
Water	= 40 ml
2.0 M DL-Malic acid, pH 8.0	= 5 ml
1.0 M $MgCl_2$	= 10 mg
$NADP^+$	= 10 ml
MTT or NBT	= 10 mg
PMS	= 1 mg

INSTRUCTIONS: Stain in the dark at room temperature. Rinse and fix.

SOURCE: Richmond (1972); Soltis et al. (1983).

Example 2

BUFFER: 0.05 M tris, pH 8.0	= 50 ml
Malic acid (pH = 8.0)	= 50 mg
$MgCl_2$	= 50 mg
$NADP^+$	= 15 mg
NBT	= 10 mg
PMS	= 1 mg

INSTRUCTIONS: Stain in the dark at 37°C. Rinse and fix.

SOURCE: Cardy and Beversdorf (1984).

Example 3

BUFFER: 0.1 M tris-malate, pH 7.2	= 50 ml
0.1 M $MgCl_2$	= 5 ml
L-Malate	= 20 mg
$NADP^+$	= 6 mg
MTT	= 6 mg
Meldola's blue	= trace

INSTRUCTIONS: Incubate in the dark at 37°C for 1 to 2 hr. Rinse and fix.

SOURCE: Weeden and Emmo (nd)

10.4.37. Mannose-6-phosphate Isomerase (MPI) E.C. 5.3.1.8

The enzyme converts mannose-6-phosphate to fructose-6-phosphate for glycolysis (Lehninger 1982).

QUATERNARY STRUCTURE: Monomer

REACTION:

$$\text{M-6-P} \xrightarrow{\text{MPI}} \text{F6P}$$

$$\text{F6P} \xrightarrow{\text{PGI}} \text{G6P}$$

$$\text{G6P} + \text{NADP}^+ \xrightarrow{\text{G6PD}} \text{6PGA} + \text{NADPH}$$

STAINING SYSTEM: NADPH is used in the tetrazolium staining system.

Example 1

BUFFER: 0.2 M tris-HCl, pH 8.0	= 10 ml
Water	= 40 ml
Mannose-6-phosphate (disodium)	= 50 mg
Glucose (or hexose) phosphate isomerase	= 100 units
Glucose-6-phosphate dehydrogenase	= 80 units
$MgCl_2$	= 50 mg
$NADP^+$	= 15 mg
NBT	= 15 mg
PMS	= 1 mg

SOURCE: Nichols et al. (1973).

Example 2

BUFFER: 0.2 M tris, pH 8.0	= 50 ml
Mannose-6-phosphate	= 20 mg

PVP-40 = 300 mg
NAD$^+$ = 10 mg
Glucose-6-phosphate dehydrogenase = 25 units
Phosphohexose (phosphoglucose) isomerase = 100 units
MTT = 5 mg
PMS = 1 mg

INSTRUCTIONS: Stain in the dark at room temperature. Rinse and fix.

SOURCE: Cardy and Beversdorf (1984).

Example 3

Agar overlay

Solution A:

BUFFER: 0.2 M tris-HCl, pH 8.0 = 15 ml
Mannose-6-phosphate = 20 mg
PVP-4 = 300 mg
NAD$^+$ = 10 mg
Glucose-6-phosphate dehydrogenase = 25 units
Phosphohexose (phosphoglucose) isomerase = 100 units
MTT = 5 mg

Solution B:

BUFFER: Same as for Solution A. = 15 ml
Agar = 200 mg

INSTRUCTIONS: Dissolve the agar by heating and cool it to 72°C before adding Solution A. Pour the mixture onto the gel and incubate for 60 min.

SOURCE: Cardy and Beversdorf (1984).

10.4.38. Nitrate Reductase (NR) E.C. 1.6.6.2

REACTION:

$$NADH + Nitrate \longrightarrow NAD^+ + Nitrite + H_2O$$

The nitrite reacts with an aryl amide to produce a dizonium.

Example

Solution 1:

BUFFER: 0.1 M Potassium phosphate, pH 7.5 = 50 ml
KNO_3 = 75 mg
95% Ethanol = 1.25 ml
NADH = 15 mg
ADH (yeast enzyme) = 50 units
(add just before incubation)

Solution 2:

1% Sulfanylamide (in 1 *N*-HCl) = 25 ml
N-1-Naphthylethylenediamine.2HCl = 25 ml
(0.01% in PO_4 buffer, pH 7.5)

INSTRUCTIONS: Incubate in Solution 1 for 30 min at 30°C. Discard the solution, rinse, and add Solution 2. Leave for a few minutes.

SOURCE: Upcroft and Done (1974).

10.4.39. Octanol Dehydrogenase (ODH) E.C. 1.1.1.73

QUATERNARY STRUCTURE: Dimer.

Example

BUFFER: 0.01 mM tris, pH 8.5 = 50 ml
2-Octanol (in ethanol 1:4) = 0.5 ml
NAD$^+$ = 75 mg
MTT = 5 mg
PMS = 1.5 mg

Caution: The prolonged staining time required may cause LDH and ADH bands to appear on the same gel stained for ODH.

SOURCE: Original source not found.

10.4.40. Pepsinogen (PEPS)

Example

BUFFER: 0.1 M Phosphate, pH 7.0
KH_2PO_4 = 13.6 g
1.0 M NaOH = 59 ml
Water = 1 l

Staining solution:
0.1 M HCl = 100 ml
Amido black dye = 1 g
Buffer = 100 ml
Bovine albumin = 0.5 g

SOURCE: Hanley et al. (1966); Shaw and Prasad (1970).

10.4.41. Peptidase (PEP) E.C. 3.4.11.— or 3.4.13.—

There are a variety of petidases, and they require various substrates:

PEP A, valine-leucine
PEP B, leucine-glycine-glycine
PEP C, lysine-leucine
PEP D, phenylalanine-proline.

Strictly, leucine aminopeptidase (LAP), described above, should be classified as a peptidase, but it is considered separately because it is one of the more frequently assayed peptidases.

QUATERNARY STRUCTURE: Monomer, dimer.

SUBCELLULAR LOCATION: Cytoplasm (Vodkin and Scandalios 1981).

REACTION:

$$\text{Peptide} \xrightarrow{\text{PEP}} \text{L-amino acid}$$

$$\text{L-amino acid} \xrightarrow{\text{L-amino acid oxidase}} H_2O_2$$

$$H_2O_2 \xrightarrow{\text{peroxidase}} O_2$$

STAINING SYSTEM: O_2 oxidizes e.g. *o*-dianisidine.

Example 1: *Peptidase E.C. 3.4.11*

BUFFER: 0.1 M Phosphate, pH 7.5 = 50 ml
Glycyl-L-leucine = 10 mg
L-Amino acid oxidase = 5 mg
Peroxidase = 10 mg
$MnCl_2.4H_2O$ (1.98 g/ 100 ml) = 0.5 ml
O-dianisidine = 5 mg

INSTRUCTIONS: Soak the gel slice in HCl for 2 hr (to lower pH to about 2.0 so as to convert pepsinogen to pepsin). Pour off Solution A and add Solution B; incubate at 37°C. Gel develops light blue background with white zones of proteolytic activity. Wash and fix.

SOURCE: Lewis and Harris (1967); Shaw and Prasad (1970).

Example 2: Peptidase E.C. 3.4.13.1

BUFFER: 0.2 M tris HCl, pH 8.0	= 50 ml
Glycyl-L-leucine	= 7.5 mg
L-Leucyl-L-tyrosine	= 7.5 mg
L-Valyl-L-leucine	= 7.5 mg
Peroxidase	= 7.5 mg
Snake venom	= 7.5 mg
3-Amino-9-ethyl carbazole	= 7.5 mg

(dissolve in 5 ml *N,N'*-dimethyl formamide and add just before staining).

SOURCE: Nichols and Ruddle (1973).

10.4.42. Peroxidase (PRX) E.C. 1.11.1.7

Peroxidases are heme enzymes whose function includes roles in fatty acid oxidation, lignification, respiration, and breakdown of indole acetic acid (Manicol 1966). In soybean, Buttery and Buzzell (1968) classified cultivars on the basis of seed coat peroxidase activity. In maize, tissue-specificity of most peroxidases was reported by Brewbaker and Hasegawa (1975). Tyson et al. (1986) reported that peroxidases are encoded by many different loci in plants and that the enzymes are subject to post-translational modification by glycosylation.

QUATERNARY STRUCTURE: Monomer (Brewbaker and Hasegawa 1975).

SUBCELLULAR LOCATION: Cytosol, cell wall (Brewbaker and Hasegawa 1975).

REACTION:

$$2H_2O_2 \xrightarrow{\text{PRX}} 2H_2O + O_2$$

Example 1

Solution A:

BUFFER: 0.05 M Na acetate, pH 5.0	= 46.25 ml
0.1 M $CaCl_2$	= 1 ml
3% H_2O_2	= 0.25 ml
3-Amino 9-ethyl carbazole (dissolve in 5 ml dimethyl formamide)	= 25 mg

Solution B:

Glycerine	= 25 ml
Water	= 25 ml

INSTRUCTIONS: Incubate gel in Solution A in a cold room (or refrigerator) until the peroxidase area is stained reddish brown (in about 30–60 min). Rinse and fix in Solution B.

SOURCE: Graham et al. (1965); Shaw and Prasad (1970).

Example 2

Spot test

Solution A:

0.5% Guaiacol

Solution B:

0.1% H_2O_2

INSTRUCTIONS: Soak a piece of dry seed coat in 10 drops of Solution A for 10 min. Add one drop of Solution B. Record the color after 20 to 40 sec. The presence of high peroxidase produces a reddish color; otherwise, no color change (i.e. this is a color–no-color test).

SOURCE: Buttery and Buzzell (1968).

10.4.43. 2,3-Phosphoglycerate Mutase (PGAM) E.C. 5.4.2.1.

Example 1

BUFFER: 0.2 M tris-HCl, pH 8.0	= 1 ml
Water	= 4 ml
Phosphoglycerate kinase	= 320 units
Glyceraldehyde-3-phosphate dehydrogenase	= 100 units
2-Phospho-glyceric acid (Na salt)	= 12.5 mg
NADH	= 15 mg
ATP	= 10 mg
$MgCl_2$	= 20 mg
EDTA	= 1 mg

INSTRUCTIONS: Apply drops of staining solution to the gel surface via a Pasteur pipette. View under long-wave u.v. light and score immediately.

SOURCE: Omenn and Cohen (1971).

Example 2

Stain buffer: 0.433 M Glycine, pH 9.0	= 50 ml
Mg acetate.$4H_2O$	= 142.5 g
Phosphoenol pyruvate (Na_3 salt)	= 14 mg
ADP	= 70 mg
LDH	= 40 units
NADH	= 56.5 mg

INSTRUCTIONS: Dissolve agar in 40 ml of glycine buffer by heating. Cool to 45°C and mix the other ingredients in 10 ml of the buffer. Add the mixture to the cooled agar solution and pour over the gel. Incubate at 37°C for 1–2 hr and view under longwave u.v. light.

10.4.44. Phosphoglucoisomerase (PGI) E.C. 5.1.3.9 (Also known as glucosephosphate isomerase or phosphohexose isomerase)

PGI functions in glycolysis (Freeling 1983); its reverse reaction is involved in gluconeogenesis (Lehninger 1982). Gottlieb (1981) reported that PGI isozymes are usually encoded by two loci.

QUATERNARY STRUCTURE: Dimer (Gottlieb 1981).

SUBCELLULAR LOCATION: Cytoplasm, plastids (Gottlieb 1981).

REACTION:

$$\text{G-1-P} \xrightarrow{\text{PGM}} \text{G-6-P}$$

$$\text{G-6-P} + \text{NADP}^+ \xrightarrow{\text{G-6PD}} \text{6-PG} + \text{NADH}$$

Example 1

BUFFER: 0.1 M tris-HCl, pH 8.0	= 50 ml
$NADP^+$	= 5 mg
$MgCl_2$	= 40 mg
Glucose-6-phosphate dehydrogenase	= 2.5 μl
Fructose 6-phosphate	= 80 mg
MTT	= 5 mg
PMS	= 0.5 mg

INSTRUCTIONS: Incubate in the dark at 37°C. Rinse and stain.
SOURCE: Delorenzo and Ruddle (1969).

Example 2

BUFFER: 0.05 M tris-HCl, pH 8.0 = 50 ml
Fructose-6-phosphate = 50 mg
$MgCl_2$ = 50 mg
NAD^+ = 10 mg
MTT = 5 mg
PMS = 1.5 mg
Glucose-6-phosphate dehydrogenase = 10 units

INSTRUCTIONS: Incubate the gel for 60 min. Rinse and fix.
SOURCE: Cardy and Beversdorf (1984).

10.4.45. Phosphoglucomutase (PGM) E.C. 2.7.5.1

This enzyme catalyzes the reversible reaction whose forward process is the breakdown of starch reserves in catabolism and whose reverse reaction has a regulatory role in gluconeogenesis in plants (Lehninger 1982). In several plant species, PGM is encoded by two loci (Goodman and Stuber 1983).
QUATERNARY STRUCTURE: Monomer (Goodman and Stuber 1983).
SUBCELLULAR LOCATION: Chloroplast, cytosol (Gottlieb 1981).
REACTION:

$$\text{G-1-P} \xrightarrow{\text{PGM}} \text{G-6-P}$$

$$\text{G-6-P} + \text{NADP}^+ \xrightarrow{\text{G-6PD}} \text{6-PG} + \text{NADPH}$$

STAINING SYSTEM: The NADPH is used in the tetrazolium staining system.

Example 1

BUFFER: 0.5 M tris-HCl, pH 7.1 = 5 ml
Water = 45 ml
Glucose 1-phosphate (disodium salt) = 300 mg
$MgCl_2.6H_2O$ = 100 mg
Glucose 6-phosphate dehydrogenase = 40 units
$NADP^+$ = 5 mg
PMS = 0.5 mg
NBT = 10 mg

INSTRUCTIONS: Incubate the gel in the dark at 37°C. Rinse and fix.
SOURCE: Spencer et al. (1964).

Example 2

BUFFER: 0.1 M tris-HCl, pH 8.5 = 50 ml
Glucose-1-phosphate = 250 mg
EDTA = 10 mg
$MgCl_2$ = 100 mg
NAD^+ = 10 mg
PMS = 1 mg
MTT = 50 mg
Glucose-6-phosphate dehydrogenase = 37.5 units

INSTRUCTIONS: Incubate the gel at room temperature for 60 min. Rinse and fix.
SOURCE: Cardy and Beversdorf (1984).

Example 3

BUFFER: 0.1 M tris-HCl, pH 8.5 = 50 ml
$MgCl_2$ = 100 mg
EDTA = 50 mg
α-D-Glucose-1-phosphate = 250 mg
Glucose-6-phosphate dehydrogenase = 40 units
$NADP^+$ = 10 mg
NBT = 10 mg
PMS = 1 mg

INSTRUCTIONS: Incubate the gel in the dark at 37°C. Rinse and fix.
SOURCE: Original source not found.

10.4.46. 6-Phosphogluconate Dehydrogenase (6-PGD) E.C. 1.1.1.44

This enzyme operates in the pentose phosphate pathway, where it catalyzes the breakdown of 6-phosphogluconate to ribulose-5-phosphate and carbon dioxide. It is usually encoded by two loci: the product of one locus is active in the cytoplasm whereas that of the other is active in plastids (Gottlieb 1981). Goodman and Stuber (1983) noted that obtaining good resolution of 6-PGD bands in maize requires caution in controlling electrophoretic conditions; for example, certain starch lots produce only streaks.
QUATERNARY STRUCTURE: Dimer (Gottlieb 1981).
SUBCELLULAR LOCATION: Cytosol, plastids (Gottlieb 1981).
REACTION:

$$\text{6-PG} + \text{NADP}^+ \xrightarrow{\text{6-PGD}} \text{CO}_2 + \text{Ribose-5-P} + \text{NADPH}$$

STAINING SYSTEM: The NADPH is used in the tetrazolium staining system.

Εχαμπλε +

BUFFER: 0.1 M tris, pH 7.5 = 50 ml
$MgCl_2$ = 49 mg
6-Phosphogluconic acid = 10 mg
$NADP^+$ = 7.5 mg
MTT = 10 mg
PMS = 2 mg

INSTRUCTIONS: Incubate the gel in the dark at 30°C. Rinse and fix.
SOURCE: Sing and Brewer (1969).

Example 2

BUFFER: 0.5 M tris-HCl, pH 7.1 = 5 ml
Water = 45 ml
6-Phosphogluconate (trisodium salt) = 100 mg
$NADP^+$ = 10 mg
NBT = 12.5 mg
PMS = 1 mg

INSTRUCTIONS: Incubate the gel in the dark at 37°C. Rinse and fix.
SOURCE: Shaw and Prasad (1970).

10.4.47. Pyruvate kinase (PK) E.C. 2.7.1.40

Example 1

BUFFER: 0.2 M tris-HCl, pH 8.0	= 1 ml
Water	= 4 ml
Lactate dehydrogenase	= 150 units
Phospho(enol) pyruvate (trisodium salt)	= 25 mg
ADP	= 2.5 mg
$MgCl_2$	= 5 mg
NADH	= 15 mg
Fructose-1,6-diphosphate	= 1 mg
EDTA	= 1 mg
KCl	= 25 mg

INSTRUCTIONS: Apply drops of staining solution to the gel surface with a Pasteur pipette. View under long-wave u.v. light and score PK bands (defluoresced) immediately.

SOURCE: Omenn and Cohen (1971).

10.4.48. Ribonuclease (RIB) E.C. 3.1.27.—

Example

BUFFER: 0.05 M Acetate, pH 5.0	= 100 ml
Yeast RNA	= 250 mg
Black K salt	= 100 mg
Acid phosphatase	= 10 mg

INSTRUCTIONS: Incubate the gel at 37°C until blue bands appear.

SOURCE: Ressler et al. (1966); Shaw and Prasad (1970).

10.4.49. Shikimate Dehydrogenase (SKDH) E.C. 1.1.1.25

Shikimic acid dehydrogenase operates in the shikimic acid pathway, the pathway by which the aromatic amino acids phenylalanine, tyrosine, and tryptophan are formed (Zubay 1983).

QUATERNARY STRUCTURE: Monomer (Moran and Bell 1983).

SUBCELLULAR LOCATION: Cytosol, plastids (Weeden and Gottlieb 1980).

REACTION:

$$\text{Shikimate} + NADP^+ \xrightarrow{\text{SKDH}} \text{Dehydro-shikimate} + NADPH$$

STAINING SYSTEM: The NADPH is used in the tetrazolium staining system.

Example 1

BUFFER: 0.1 M tris, pH 7.5	= 50 ml
Shikimic acid	= 50 mg
$NADP^+$	= 7.5 mg
MTT	= 10 mg
PMS	= 2 mg

INSTRUCTIONS: Incubate the gel in the dark at 30°C. Rinse and fix.

SOURCE: Tanksley and Rick (1980).

Example 2

NBT may be substituted for PMS (Soltis et al. 1983).

10.4.50. Sorbitol Dehydrogenase (SDH) E.C. 1.1.1.14

Example 1

BUFFER: 0.05 M tris-HCl, pH 8.0	= 50 ml
Sorbitol	= 250 mg
NAD^+	= 5 mg
MTT	= 7.5 mg
PMS	= 1 mg

INSTRUCTIONS: Incubate the gel at 37°C. Rinse and fix.

SOURCE: Lin et al. (1969).

Example 2

BUFFER: 0.2 M tris-HCl, pH 8.0	= 50 ml
Sorbitol	= 2 g
NAD^+	= 10 mg
MTT	= 10 mg
PMS	= 1 mg

SOURCE: O'Malley et al. (1980).

10.4.51. Succinate Dehydrogenase (SUD) E.C. 1.3.99.1

This TCA cycle enzyme catalyzes the reaction of the conversion of succinate to fumarate.

REACTION:

$$\text{Succinate} + NAD^+ \xrightarrow{\text{SDH}} \text{fumarate} + NADH$$

STAINING SYSTEM: The NADH is used in the tetrazolium staining system.

Example

BUFFER: 50 mM Na phosphate, pH 7	= 50 ml
Na EDTA	= 200 mg
Na Succinate	= 125 mg
Na ATP	= 25 mg
NAD^+	= 35 mg
NBT	= 20 mg
PMS	= 1 mg

INSTRUCTIONS: Incubate the gel at 30°C for 30 min.

SOURCE: Brewer and Sing (1970).

10.4.52. Superoxide Dismutase (SOD) E.C. 1.15.1.1 [Also called tetrazolium oxidase (TO)]

Superoxide dismutase is ubiquitous among oxygen-metabolizing organisms. It catalyzes the reaction of an exchange of a single electron from one substrate molecule to another (McCord 1979). It is visible as a negative stain (achromatic). Its physiological function may be to protect against metabolically generated superoxide radicals (McCord et al. 1971).

QUATERNARY STRUCTURE: Dimer, tetramer (Baum and Scandalios 1981).

SUBCELLULAR LOCATION: Cytosol, plastid (Baum and Scandalios 1981).

REACTION:

$$O_2^- + O_2^- + 2H^+ \xrightarrow{\text{SOD}} O^2 + H_2O_2$$

Example 1
Solution A:
BUFFER: Na phosphate 50 mM, pH 7.5 = 50 ml
MTT = 100 mg
Solution B:
BUFFER: Same as for Solution A = 50 ml
TEMED = 0.2 ml
Riboflavin = 0.5 mg
INSTRUCTIONS: Soak the gel in Solution A for 20 min in the dark at 30°C. Drain and add Solution B and incubate under light.
SOURCE: Beauchamp and Fridovich (1971).

Example 2
BUFFER: 0.05 M tris-HCl, pH 8.0 = 50 ml
$MgCl_2$ = 1 mg
$NADP^+$ = 10 mg
NBT = 10 mg
PMS = 0.5 mg
INSTRUCTIONS: Incubate the gel for 60 min. Rinse and fix in water. Bands appear overnight.
SOURCE: Cardy and Beversdorf (1984).

10.4.53. Triosephosphate isomerase (TPI) E.C. 5.3.1.1

Triosephosphate isomerase catalyzes the interconversion between glyceraldehyde-3-phosphate and DHAP (Zubay 1983).
QUATERNARY STRUCTURE: Dimer (Tanksley and Rick 1980).
REACTION:

$$\text{DHAP} \xrightarrow{\text{TPI}} \text{GA-3-P}$$

$$\text{GA-3-P} + \text{NAD} + \text{phosphate/arsenate} \xrightarrow{\text{GA-3-PD}} \text{1,3-DPG} + \text{NADH}$$

Example 1
BUFFER: 1.0 M tris-HCl, pH 8.0 = 5 ml
Water = 45 ml
Dihydroxyacetone phosphate (DHAP) (lithium salt) = 5 mg
EDTA (tetrasodium salt, dihydrate) = 19 mg
NAD^+ = 15 mg
MTT = 5 mg
PMS = 1 mg
Arsenic acid (sodium salt) = 230 mg
Glyceraldehyde-3-phosphate dehydrogenase = 150 units
INSTRUCTIONS: Stain in the dark at 37°C.
SOURCE: Modified by Soltis et al. (1983).

Example 2
Substrate: Dihydroxyacetone phosphate
BUFFER: 0.2 M tris-HCl, pH 8.0 = 10 ml
Water = 70 ml
1 M Na-α-Glycerophosphate = 10 ml
1 M Na pyruvate = 10 ml
NAD^+ = 50 mg
α-Glycerophosphate dehydrogenase = 200 μg
L-lactic acid dehydrogenase = 200 units
INSTRUCTIONS: Incubate at 37°C for 2 hr. Adjust the pH to 2.0 with 1 *N* HCl in order to inactivate the enzyme. Readjust the pH to 7.0 with 1 M tris.
Staining solution:
NAD^+ = 30 mg
NBT = 15 mg
PMS = 1 mg
Arsenic acid (sodium salt) = 125 mg
Substrate solution = 50 ml
Phosphoglyceraldehyde dehydrogenase = 5 mg
INSTRUCTIONS: Incubate the gel at 37°C. Rinse and fix.
SOURCE: Scopes 1968 (cited by Shaw and Koen 1968); Shaw and Prasad 1970.

10.4.54. Xanthine dehydrogenase (XDH) E.C. 1.2.1.37

REACTION:

$$\text{Hypoxanthine} + NAD^+ + H_2O \longrightarrow \text{xanthine} + \text{NADH}$$

$$\text{Xanthine} + NAD^+ + H_2O \longrightarrow \text{Urate} + \text{NADH}$$

STAINING SYSTEM: NADH is used in the tetrazolium staining system.

Example
BUFFER: 50 mM tris, pH 7.5 = 50 ml
Hypoxanthine = 350 mg
NAD^+ = 25 mg
NBT = 10 mg
PMS = 2 mg
INSTRUCTIONS: Heat to dissolve the hypoxanthine in buffer. Cool and add other reagents. Incubate the gel in dark the at 30°C for 30–60 min. Rinse and fix.
SOURCE: Brewer and Sing (1970).

10.4.55. General Proteins

Example 1
Naphthol blue black solution: May be used e.g. for phaseolin and ribulose-1,6-bisphosphate carboxylase (rubisco).
Wash solution:
Methanol = 100 ml
Water = 100 ml
Acetic acid = 20 mg
Staining solution:
Wash solution = 50 ml
Naphthol blue black = 50 mg
INSTRUCTIONS: Filter the staining solution before use. Pour over the gel and incubate for 1 hr or more at room temperature. Rinse with wash solution (about 2–3 times). Fix in wash solution.
SOURCE: Weeden and Emmo (nd).

Example 2
Coomassie brilliant blue solution: May be used e.g. for general staining of blood samples.
Coomassie BB-250 = 5 g
Methanol = 225 ml
Water = 225 ml
Acetic acid = 50 ml
SOURCE: Original source not found.

GLOSSARY

Allele: One of two or more alternative forms of a gene located at a single locus. A *diploid* has two alleles, a *triploid* has three, and so forth. Each allele has a specific and unique nucleotide sequence. Alleles are distinguished on the basis of their effects on phenotype. In isozyme studies, a phenotype is a specific electrophoretic pattern.

Allele frequency: The proportion of all alleles of a gene in a population that are of a specified type.

Allopolyploid: An organism whose genome is comprised of two different chromosome sets (from different species). Union of more than two different chromosome sets can occur, as in wheat.

Allozymes: Isozymes that are encoded by different allelic genes. They are distributed in a population according to Mendelian laws.

Amino acids: Compounds that constitute the basic building blocks of proteins.

Autopolyploid: An organism whose genome is comprised of identical chromosome sets from the same species. An organism with two identical sets of chromosomes is called a *diploid*.

Band: A spot (chromatic or achromatic) on a stained gel indicating the position of enzyme activity.

Buffer tank/tray: Receptacle that holds the electrolyte solution used during electrophoresis.

Codominant alleles: Alleles whose phenotypes are expressed simultaneously in a heterozygote. Isozyme alleles usually exhibit codominance of expression.

Comigration: The movement of different protein molecules in electrophoresis to the same spot on a stained gel.

Cryptic variation: In the context of electrophoresis, cryptic variation is variation not measured by the technique (e.g. charges of molecules that do not contribute to electrophoretic mobility because they are "hidden" by the quaternary structure of the protein, or variants carry the same charge; also, variability in non-coding parts of DNA are not available to electrophoretic technique).

Degas (a gel): To aspirate a gel mixture in order to rid it of air and bubbles before it solidifies.

Denatured protein: Protein that has been treated by agents such as heat so that its native configuration has been lost (e.g. polypeptides are dissociated).

Destain: To remove background stain from a gel to expose the desired bands for easy scoring.

Develop: To render bands ("images") visible during the staining of a gel.

Dewick: To remove wicks from a gel after a short period of electrophoresis.

Diploid: An organism (or cell, tissue) with two homologous chromosome sets.

Dominant allele: An allele whose phenotypic effect is expressed to the exclusion of the effect of an alternative allele (recessive) in a heterozygote.

Duplication: A chromosome mutation in which a segment of a chromosome is repeated in the haploid genome. Duplication is a known origin of isozymes.

Electromorphs: Isozymes that are distinguishable on the basis of electrophoretic results. These distinctly different patterns on a zymogram are also called *zymotypes, electrophoretic variants,* or *variant patterns.*

Electrophoresis: A technique for separating a mixture of charged particles in a suitable medium placed in an electric field. This is one of the major techniques of isozyme research and precedes enzyme activity staining for visualization of isozymes.

Enzyme: A protein with a catalytic activity. Isozymes are multiple molecular forms of an enzyme.

Enzyme Commission number (E.C.): Usually a four-section number separated by periods, each section of the number corresponding to the main class, subclass, sub-subclass, and serial number of an enzyme, respectively.

Epigenetic change: An alteration in phenotypic expression that is not a consequence of a genotypic change. Such changes may produce isozymes of varying stability and mobility.

Epistasis: An interlocus gene interaction in which the expression of one allele precludes the phenotypic effects of a nonallelic gene.

Fingerprint: A pattern of bands or spots visible on a zymogram and characterizing a specific enzyme, genotype, and electrophoretic condition. Where more than one enzyme is involved, the fingerprint refers to the composite of all the enzymes used to identify the genotype.

Fix: To use the appropriate mixture of reagents to arrest the staining process and to preserve the colored bands for a period of time.

Genotype: The collective genetic information within an organism. It is also used to describe the genic constitution of an individual with respect to one or more gene loci of interest.

Heterozygosity: On an individual basis, heterozygosity refers to the proportion of heterozygous loci an individual has; on a population basis, it refers to the proportion of heterozygous individuals at a locus.

Heterozygote: An individual or a cell that has two alleles at a given locus on homologous chromosomes.

Homozygote: An individual or cell that has identical alleles at a given locus on homologous chromosomes.

Horizontal gel: An electrophoretic support medium cast so that electrophoresis is conducted while the gel is in a horizontal position.

Isozymes: Multiple molecular forms of an enzyme in a species that share a common catalytic activity. The different forms appear as electrophoretic variants upon staining of a gel for the detection of a specific enzyme.

Linkage: Tendency of genes to be inherited together by virtue of their being located on the same chromosome. Genes located closely together are less likely to be separated by recombination.

Load (a gel) : To insert specimen wicks (or place liquid specimens) into a gel for electrophoresis.

Locus: Frequently used synonymously with gene, locus refers to the specific location on a genetic map at which a gene is found.

Marker/tracking dye: A dye that is loaded along with specimens in a gel to indicate the direction and distance of migration in electrophoresis. The migration distance of the marker is used as a reference point in determining the relative mobility of other molecules.

Mold: The apparatus in which a fluid gel mixture is poured for solidification. Electrophoresis is conducted while the gel is in the mold.

Monomer: Consists of one subunit. A monomeric enzyme or protein has a single subunit encoded by a single polypeptide.

Multimer: Consists of more than one subunit. A multimeric enzyme or protein has two or more subunits, each encoded by a different polypeptide. A *dimer* consists of two subunits, a *trimer* of three, a *tetramer* of four, and so forth. If the subunits are identical, the multimer is called a *homomer* or *homomultimer* (e.g. *homodimer*), otherwise it is called a *heteromer* or *heteromultimer* (e.g. *heterodimer*).

Neutrality: Alleles are said to be selectively neutral when the different alleles at a locus confer equal fitness (no selective advantage) to the organism.

Null allele: An allele whose presence is manifested by a lack of a normal gene product or loss of a specific function. In electrophoretic analysis, a null allele is recorded when a band or spot is missing on a stained gel as compared to a reference pattern. Null alleles must always be authenticated with segregation analysis, since nonideal electrophoretic conditions can hinder the full expression of all the possible bands.

Origin: The starting or loading point from which protein samples migrate through an electrophoretic support medium. This may consist of a slit in the gel into which specimen wicks are loaded or indentations into which liquid specimens are placed.

PAGE: Polyacrylamide gel electrophoresis. Electrophoresis conducted with a polyacrylamide gel as the support medium.

Phenotype: The product of the interaction between a genotype and its environment. Fingerprints produced from electrophoresis represent specific phenotypes.

Ploidy: The number of sets of chromosomes in a cell or an organism. A diploid has two sets of chromosomes, a triploid has three, and so forth.

Polymorphism: The simultaneous occurrence of several forms of a trait or gene in a population or between populations. Isozyme polymorphisms are ubiquitous in organisms and are used as markers in a variety of applications.

Pour (a gel) : To deliver the fluid mixture of gel constituents (starch, polyacrylamide, etc.) into a mold (usually refers to all the steps involved in casting or molding a gel for electrophoresis).

Primary structure of a protein: Amino acids in the polypeptide chain.

Quaternary structure of a protein: The multimeric constitution of the protein.

Read (score) a gel: To gather information from electrophoretic patterns of a stained gel.

Recessive allele: An allele whose phenotypic expression is suppressed by the presence of a dominant allele in a heterozygote but is observed in a homozygous state.

Relative mobility: The location of a spot or band on a zymogram in relation to the location of a marker dye or other protein.

Resolution: The band definition (width and sharpness).

Run a gel: To place a gel loaded with specimens in an electric field, or simply to conduct electrophoresis.

Separation: The distance between bands on a zymogram.

SGE: Starch gel electrophoresis. Electrophoresis conducted with starch as the medium of support.

Slow/fast band: Designations used by some researchers to describe the relative migration of protein molecules in a gel. A fast band is located further away from the origin than a slow band.

Sponge: The absorbent material used to link an electrolyte with a gel. (See also Wick.)

Structural gene: Codes for the amino acid sequence of a polypeptide.

Subunit: One polypeptide chain in a protein.

Tertiary structure of a protein: The folding of the secondary structure to produce globular molecules.

Vertical gel: An electrophoretic support medium that is cast so that electrophoresis is conducted while the gel is vertically positioned.

Wicks: Narrow strips of absorbent paper (e.g. Whatman paper no. 3) used to soak up sample extract for electrophoresis. The absorbent material used to link the electrode buffer with the support medium in the Type I electrophoresis unit is referred to as a wick.

LITERATURE CITED

Afanador, L. 1989. *Characterization of Two Bean Fungi by Isozyme Analysis.* M.S. Thesis. Michigan State University.

Aicher, D. L. 1988. *Investigations of Isozymes in Sugar Beet* (Beta vulgaris L.). Ph.D. Thesis. Michigan State University.

Akroyd, P. 1965. Acrylamide gel electrophoresis of β-lactoglobulins stored in solutions of pH 8.7. *Nature* 208:488–489.

Allard, R. W. 1956. Formulas and tables to facilitate the calculations of recombination values in heredity. *Hilgardia* 24:235–278.

Allard, R. W., G. R. Babbel, M. T. Clegg, and A. L. Kahler. 1972. Evidence for coadaptation in *Avena barbata. Proc. Nat. Acad. Sci. USA* 69:3043–3048.

Allen, S. L., M. S. Misch, and B. M. Morrison. 1963. Acid phosphatase of *Tetrahymena. J. Histochem. Cytochem.* 11:706–711.

Allendorf, F. W. 1977. Electromorphs or alleles. *Genetics* 87:821–822.

An Der Lan, B., and A. Chrambach. 1981. Analytical and preparative gel electrophoresis. In *Gel Electrophoresis of Proteins a Practical Approach.* Eds. B. D. Hames and R. Rickwood. Oxford, UK: IRL Press. 157–187.

Anderson, J. W. 1968. Extraction of enzymes and subcelluar organelles from plant tissues. *Phytochemistry* 7:1973–1988.

Arulsekar, S., and R. S. Bringhurst. 1981. Genetic model for the enzyme maker PGI, its variability and use in accelerating the mating system in diploid California *Fragaria vesca* L. *J. Hered.* 72:117–120.

Arus, P. 1983. Genetic purity of commercial seed lots. In *Isozymes in Plant Breeding and Research,* Part A. Eds. S. D. Tanksley and T. J. Orton. Amsterdam: Elsevier. 415–423.

Arus, P., and J. J. Orton. 1983. *J. Hered.* 74:405–412.

Aston, C. C., and A. W. H. Braden. 1961. Serum β-globulin polymorphism in mice. *Aust. J. Biol. Sci.* 14:248–253.

Ayala, F. J., and J. A. Kiger. 1980. Gel electrophoresis. In *Modern Genetics.* Menlo Park, CA: Benjamin/Cummings.

Babson, A. L., P. O. Shapiro, P. A. R. Williams, and G. E. Philips. 1962. The use of a diazonium salt for the determination of glutamic oxalacetic transaminase in serum. *Clin. Chim. Acta* 7:199–205.

Basssiri, A., and I. Rouhani. 1971. Identification of broadbean cultivars based on isozyme patterns. *Euphytica* 26:279–286.

Baum, J. A., and J. G. Scandalios. 1979. Developmental expression and intracellular location of superoxide dismutase in maize. *Differentiation* 13:133–140.

______. 1981. Isolation and characterization of the cytosolic and mitochondrial superoxide dismutase of maize. *Archives Biochem. Biophys.* 206:249–264.

______. 1982. Multiple genes controlling superoxide dismutase expression in maize. *J. Hered.* 73:95–100.

Beauchamp, C., and J. Fridovich. 1971. Superoxide dismutase: Improved assays and an assay applicable to polyacrylamide gels. *Anal. Biochem.* 44:276–287.

Bingham, E. T. 1980. Maximizing the heterozygosity in autopolyploids. In *Polyploid Biological Relevance.* Ed. W. H. Lewis. New York: Plenum. 471–490.

Boyer, S. H. 1961. Alkaline phosphatase in human serum and placenta. *Science* 134:1002.

Bradford, M. M. 1976. A rapid and sensitive method for the quantitation of microgram quantities of proteins using the principle of protein-dye binding. *Anal. Biochem.* 72:248–254.

Brewbaker, J. L., and Y. Hasegawa. 1975. Polymorphism of the major peroxidases in maize. In *Isozymes III, Developmental Biology.* Ed. C. L. Markert. New York: Academic Press. 659–673.

Brewer, G. J., J. W. Eaton, C. S. Knutsen, and C. C. Beck. 1967. A starch electrophoretic method for the study of diaphorase isozyme and preliminary results with sheep and human erythrocytes. *Biochem. Biophys. Res. Commun.* 29:198–204.

Brewer, J. B., and C. F. Sing. 1970. *An Introduction to Isozyme Techniques.* New York: Academic Press.

Bringhurst, R. S., S. Arulsekar, J. Hancock, Jr., and V. Voth. 1981. Electrophoretic characterization of strawberry cultivars. *J. Amer. Soc. Hort. Sci.* 106:684–687.

Brown, A. D. H. 1978. Isozyme, plant population, genetic structure, and genetic conservation. *Theor. Appl. Genet.* 52:145–157.

______. 1983. Barley. In *Isozymes in Plant Genetics and Breeding.* Part B. Eds. S. D. Tanksley and T. J. Orton. Amsterdam: Elsevier. 55–77.

Brown, A. D. H., J. J. Burdon, and A. M. Jarosz. 1989. Isozyme analysis of plant mating systems. In *Isozymes in Plant Biology.* Eds. D. E. Soltis and P. S. Soltis. Portland, OR: Dioscorides Press. 73–105.

Brown, A. D. H., and M. T. Clegg. 1983. Isozyme assessment

of plant genetic resources. In *Isozymes: Current Topics in Biological and Medical Research*. Eds. M. C. Rattazzi, I. G. Scandalios, and G. S. Whitt. New York: Allan R. Liss. 11:285–295.

Brown, A. D. H., D. R. Marshall, and J. Munday. 1976. Adaptedness of variation at an alcohol dehydrogenase locus in *Bromus mollis*. *Aust. J. Biol. Sci.* 29:389–396.

Brown, A. D. H., E. Nevo, D. Zohary, and O. Dagan. 1978. Genetic variation in natural populations of wild barley (*Hordeum spontaneum*). *Genetics* 49:97–108.

Brown, A. D. H., and B. S. Weir. 1983. Measuring genetic variability in plant populations. In *Isozymes in Plant Genetics and Breeding*, Part B. Eds. S. D. Tanksley and T. J. Orton. Amsterdam: Elsevier. 219–240.

Buttery, B. R., and R. I. Buzzell. 1968. Peroxide activity in seeds of soybean varieties. *Crop Sci.* 8:722–725.

Cardy, B. J., and W. D. Beversdorf. 1984. A procedure for the starch gel electrophoretic detection of isozymes of soybean (*Glycine max* L. Merr.). Department of Crop Science Tech. Bull. 119/8401. University of Guelph, Ontario, Canada.

Cardy, B. J., C. W. Stuber, and M. M. Goodman. 1981. Techniques for starch gel electrophoresis of enzymes from maize (*Zea Mays* L.). Institute of Statistics Mimeograph Series No. 1317. North Carolina State University, Raleigh.

Carr, B., and G. B. Johnson. 1980. Polyploidy, plants and electrophoresis. In *Polyploidy Biological Relevance*. Ed. W. H. Lewis. New York: Plenum. 521–528.

Catsimpoolas, N., and J. Drysdale. 1977. *Biological and Biomedical Applications of Isoelectric Focusing*. New York: Plenum.

Cavener, D., and M. T. Clegg. 1981. Evidence for biochemical and physiological differences between isozyme genotypes in *Drosophila melanogaster*. *Proc. Natl. Acad. Sci. USA* 87:4444–4447.

Cheliak, W. M., and J. A. Pitel. 1984. Techniques for starch gel electrophoresis of enzymes from forest trees. Information Report PI-X-42. Petawawa National Forestry Institute, Canadian Forest Service.

Chen, S. H., E. R. Giblett, J. E. Anderson, and B. G. L. Fossum. 1972. Genetics of glutamic-pyruvic transaminase: Its inheritance, common and rare variants, population distribution and differences in catalytic activity. *Ann. Hum. Genet. London* 35:401–409.

Chen, S. H., and H. E. Sutten. 1967. Bovine transferins: Sialic acid and the complex phenotype. *Genetics* 56:425–430.

Cherry, J. P., F. R. H. Katterman, and J. E. Endrizzi. 1970. A comparative study of seed proteins of allopolyploids of Gossypium by gel electrophoresis. *Can. J. Genet. Cytol.* 13:155–158.

Chrambach, A. 1966. Device for sectioning of polyacrylamide gel cylinders and its use in determining biological activity in the sections. *Anal. Biochem.* 15:544–548.

Chrambach, A., R. A. Reisfield, M. Wyckoff, and J. Zaccari. 1967. A procedure for rapid and sensitive staining of protein fractionated by polyacrylamide gel electrophoresis. *Anal. Biochem.* 20:150–154.

Chrambach, A., and D. Rodbard. 1971. Polyacrylamide gel electrophoresis. *Science* 172:440–451.

Clayton, J. W., and D. N. Tretiak. 1972. Amine-citrate buffers for pH control in starch gel electrophoresis. *J. Fish. Board. Can.* 29:1169–1172.

Clegg, M. T., R. W. Allard, and A. L. Kahler. 1972. Is the gene the unit of selection? Evidence from two experimental plant populations. *Proc. Nat. Acad. Sci. USA* 69:2472–2478.

Correa-Victoria, F. J. 1987. *Pathogenic Variation, Production of Toxic Metabolites and Isoenzyme Analysis in* Phaeoisariopsis griseola (*Sacc.*) *Ferr.* Ph.D. Thesis. Michigan State University.

Coyne, J. A. 1982. Gel electrophoresis and cryptic protein variation. In *Isozymes: Current Topics in Biological and Medical Research*. New York: Allan R. Liss. 6:1–32.

Crawford, D. J. 1983. Phylogenetic and systematic inferences from electrophoretic studies. In *Isozymes in Plant Genetics and Breeding*, Part A. Eds. S. D. Tanksley and T. J. Orton. Amsterdam: Elsevier. 257–287.

______. 1989. Enzyme electrophoresis and plant systematics. In *Isozymes in Plant Biology*. Eds. D. E. Soltis and P. S. Soltis. Portland OR: Dioscorides Press. 146–164.

Davis, B. J. 1964. Disc electrophoresis. II: Method and application to human serum protein. *Ann. NY Acad. Sci.* 121:404–447.

Dawson, M., H. M. Eppenbergen, and N. O. Naughton 1966. Creatine kinase: Evidence of a dimeric structure. *Biochem. Biophys. Res. Commun.* 21:346.

DeJong, D. W. 1973. Effect of temperature and day length on peroxidase and malate (NAD) dehydrogenase isozymic composition in tobacco leaf extracts. *Amer. J. Bot.* 60:846–852.

DeJong, E. M. B. 1955. Eine genaue methode zur papierelektrophoretischen auswertung von proteinen. *Rec. Trav. Chim.* 74:128.

Delorenzo, R. J., and F. H. Ruddle. 1969. Genetic control of two electrophoretic variants of glucosephosphate isomerase in the mouse (*Mus muscullus*). *Biochem. Genet.* 3:51.

deWet, J. M. J. 1980. Origin of polyploids. In *Polyploidy Biological Relevance*. Ed. W. H. Lewis. New York: Plenum. 3–16.

Dickman, S. R., and J. F. Speyer. 1954. Factors affecting the activity of mitochondrial and soluble aconitase. *J. Biol. Chem.* 206:67–75.

Diezel, W., G. Kopperschlager, and R. Hoffman. 1972. An improved procedure for protein staining in polyacrylamide gels with a new type of Coomassie Brilliant Blue. *Anal. Biochem.* 48:617–620.

Dixon, M., and E. C. Webb. 1979. *Enzymes*. London: Longmans.

Doebley, J. 1989. Isozymic evidence and the evolution of crop plants. In *Isozymes in Plant Biology*. Eds. D. E. Soltis and P. S. Soltis. Portland OR: Dioscorides Press. 165–191.

Doong, J. Y. H., and Y. T. Kiang. 1987. Inheritance of aconitase isozymes in soybean. *Genome* 29:713–717.

Dykhuizen, D., and D. L. Hartl. 1980. Selective neutrality of 6PGD allozymes in *E. coli* and the effects of genetic background. *Genetics* 96:801–817.

Eaton, G. M., G. J. Brewer, and R. E. Tashian. 1966. Hexokinase isozyme patterns of human erythrocyctes and leukocytes. *National Meeting on Hematology*. (Abstr.).

Edwards, M. D., C. W. Stuber, and J. F. Wendel. 1987. Molecular-marker-facilitated investigations of quantitative trait loci in maize: I. Number, genomic distribution and types of gene action. *Genetics* 116:113–125.

Efron, Y. 1970. Tissue specific variation in the isozyme patterns of the *Ap1*, acid phosphatase in maize. *Genetics* 65;575–583.

Endo, T., and H. Morishma. 1983. Rice. In *Isozymes in Plant Genetics and Breeding*, Part B. Eds. S. D. Tanksley and T. J. Orton. Amsterdam: Elsevier. 125–146.

Fawcett, K., and Morris, J. 1966. *Separation* Sci. 1,9.

Fildes, R. A., and H. Harris. 1966. Genetically determined variation of adenylate kinase in man. *Nature* 209:261.

Fine, I. H., and L. A. Costello. 1963. The use of starch gel electrophoresis in dehydrogenase studies. In *Research in*

Enzymology VI. Eds. S. P. Colowick and N. O. Kaplan. New York: Academic Press. 958–972.

Fondo, E. Y. Jr., and M. Bartalos. 1969. Electrophoretic separation of multiple bands with beta glucuronidase activity in human sera. *Biochem. Genet.* 3:591.

Freeling, M. 1983. Isozyme systems to study gene regulation during development: A lecture. In *Isozymes in Plant Genetics and Breeding*, Part A. Eds. S. D. Tanksley and T. J. Orton. Amsterdam: Elsevier. 61–84.

Frydenberg, O., and G. Nielsen. 1966. Amylase isozymes in germinating barley seeds. *Hereditas* 54:123–139.

Gabriel, O., and S-F. Wang 1969. Determination of enzymatic activity in polyacrylamide gel. 1: Enzymes catalyzing the conversion of nonreducing substrates to reducing susbstrates. *Anal. Biochem.* 27:545–554.

Gates, P., and D. Boulter. 1979. The use of seed isozymes as an aid to the breeding of field beans (*Vicia faba* L.) *New Phytol.* 83:783–791.

Goodman, M. M., and C. W. Stuber. 1983. Isozymes of maize. In *Isozymes in Plant Genetics and Breeding*, Part B. Eds. S. D. Tanksley and T. J. Orton. Amsterdam: Elsevier. 231–241.

Goodman, M. M., C. W. Stuber, C. M. Lee, and F. M. Johnson. 1980. Genetic control of malate dehydrogenase isozymes in maize. *Genetics* 94:153–168.

Gorman, M. B. 1983. *An Electrophoretic Analysis of the Genetic Variation in the Wild and Cultivated Soybean Germplasm*. Ph.D. Thesis, University of New Hampshire.

Gorman, M. B., and Y. T. Kiang. 1977. Variety-specific electrophoretic variants of some soybean enzymes. *Crop Sci.* 17:963–965.

______. 1978. Models for the inheritance of several variant soybean electrophoretic zymograms. *J. Hered.* 69:255–258.

Gorman, M. B., Y. T. Kiang, Y. C. Chiang, and R. G. Palmer. 1982. Preliminary electrophoresis observations from several soybean enzymes. *Soybean Genet. Newslett.* 9:140–143.

Gorovsky, M. A., K. Calson, and J. L. Rosenbaum. 1970. Simple method for quantitative densitometry of polyacrylamide gels using fast green. *Anal. Biochem.* 35:359–370.

Gottlieb, L. D. 1973. Genetic control of glutamte oxaloacetate transaminase isozymes in the diploid plant *Stephanomeria exigua* and its allotetraploid derivative. *Biochem. Genet.* 9:97–107.

______. 1974. Genetic confirmation of the origin of *Clarkia lingulat. Evolution* 28:244–250.

______. 1981. Electrophoretic evidence and plant populations. *Progr. Phytochem.* 7:2–46.

______. 1982. Conservation and duplication of isozymes in plants. *Science* 216:373–380.

Gottlieb, L. D., and N. G. Weeden. 1979. Gene duplication and phylogeny in *Clarkia. Evolution.* 33:1024–1039.

______. 1981. Correlation between subcellular location and phosphoglucose isomerase variability. *Evolution* 35:1019–1022.

Graham, R. C., H. Lundholm, and M. J. Karnovsky. 1965. Cytochemical demonstration of peroxidase activity with 3-amino-9-ethylcarbazole. *J. Histochem. Cytochem.* 13:150–152.

Griffin, J. D., and R. G. Palmer. 1987. Inheritance and linkage studies with five isozyme loci in soybean. *Crop Sci.* 27:885–892.

Hall, R. 1967. Proteins and catalase isoenzymes from *Fusarium solani* and their taxanomic significance. *Aust. J. Biol. Sci.* 20:419–428.

Hames, B. D. 1981. An introduction to polyacrylamide gel electrophoresis. In *Gel Electrophoresis of Proteins, a Practical Approach*. Eds. B. D. Hames and D. Rickwood. Oxford, UK: IRL Press. 1–99.

Hanley, W. B., S. H. Boyer, and M. A. Naughton. 1966. Electrophoretic and functional heterogeneity of pepsinogen in several species. *Nature* 209:996–1002.

Hare, R. C. 1966. Physiology of resistance to fungal disease in plants. *Bot. Rev.* 32(2):95–137.

Harris, H., and D. A. Hopkinson. 1976. *Handbook of Enzyme Electrophoresis in Human Genetics*. New York: American Elsevier.

Hart, G. E. 1969. Genetic control of alcohol dehydrogenase isozymes in *Triticum dicoccum. Biochem. Genet.* 3:617–625.

______. 1983. Hexaploid wheat (*Triticum aestivum* L. Thell). In *Isozymes in Plant Genetics and Breeding*, Part B. Eds. S. D. Tanksley and T. J. Orton. Amsterdam: Elsevier. 35–56.

Hartman, T., M. Nagel, and H. I. Ilert. 1973. Organ specific multiple forms of glutamic dehydrogenase in *Medicago sativa. Planta* 111:119–128.

Hauptli, H., and S. K. Jain. 1978. Biosystematics and agronomic potential of some weedy and cultivated amaranths. *Theor. Appl. Genet.* 52:171–185.

Henderson, N. S. 1965. Isozymes of isocitrate dehydrogenase: Subunit structure and intracelluar location. *J. Exp. Zool.* 158:263.

Herbert, M., J. Gauldiea, and B. L. Hillcoal. 1972. Multiple enzyme forms from protein-bromophenol blue interaction during gel electrophoresis. *Anal. Biochem.* 46:433–437.

Hjerten, S. 1962. 'Molecular sieve' chromatography on polyacrylamide gels prepared according to a simplified method. *Arch. Biochem. Biophys. Suppl.* 1:147.

Hjerten, S., S. Jerstedt, and A. Tiselius. 1969. Apparatus for large-scale preparative polyacrylamide gel electrophoresis. *Anal. Biochem.* 27:108–129.

Honold, G. R., G. L. Farkas, and M. A. Stahmann. 1966. The oxidation-reduction enzymes of wheat. I. A quantitative investigation of the dehydrogenases. *Cereal Chem.* 43:517–529.

Hopkinson, D. A., V. H. Edwards, and H. Harris. 1976. The distribution of subunit numbers and subunit sizes of enzymes. A study of the products of 100 human gene loci. *Ann. Hum. Genet.* 39:383–385.

Hubby, J. L., and R. C. Lewontin. 1966. A molecular approach to the study of genic heterozygosity in natural populations. I. The number of alleles at different loci in *Drosophila pseudoobscura. Genetics* 54:577–594.

Hughes, D. L. 1981. *Identification and Translocation of Carbohydrates in the Cantaloupe* (Cucumis melo *var* reticulatus) *Plant and the Fate of Stacyose During Fruit Development*. Ph.D. Thesis. University of California.

Hunter, R. L., and C. L. Markert. 1957. Histochemical demonstration of enzymes separated by zone electrophoresis in starch gels. *Science* 125:1294–1295.

Ingram, V. M. 1957. Gene mutations in human haemoglobin. The chemical difference between normal and sickle cell haemoglobin. *Nature* 180:326–328.

International Union of Biochemistry (IUB) Nomenclature Committee. 1979. *Enzyme Nomenclature* New York: Academic press.

International Union of Pure and Applied Chemistry–International Union of Biochemistry Commission on Biochemical Nomenclature. 1976. Recommendations (1976): Nomenclature of multiple forms of enzymes. *Eur. J. Biochem.* 82:1–3.

Jaaska, V. 1978. NADP-dependent aromatic alcohol dehydrogenase in polyploid wheats and their diploid relatives. On the origin and phylogeny of polyploid wheats. *Theor. Appl. Genet.* 53:209–217.

______. 1983. Secale and triticale. In *Isozymes in Plant Genetics and Breeding,* Part B. Eds. S. D. Tanksley and T. J. Orton. Amsterdam: Elsevier. 79–101.

Johnson, F. M., and H. E. Schaffer. 1974. An inexpensive apparatus for horizontal gel electrophoresis. *Isozyme Bull.* 7:4–6.

Johnson, G. B. 1974. Enzyme polymorphism and metabolism. *Science* 184:28–37.

______. 1976. Genetic polymorphism and enzyme function. In *Molecular evolution.* Ed. F. J. Ayala. Boston, MA: Sinauer Inc. 46–59.

Jordan, E. M., and S. Raymond. 1969. Gel electrophoresis: A new catalyst for acid systems. *Anal. Biochem.* 27:205–211.

Kahler, A. L., and R. W. Allard. 1970. Genetics of isozyme variants in barley. I. Esterase. *Crop Sci.* 10:444–448.

Kahler, A. L., R. W. Allard, M. Krzakowa, C. F. Wehrhahn, and E. Nevo. 1980. Associations between isozyme phenotypes and environment in the slender wild oat (*Avena barbata*) in Israel. *Theor. Appl. Genet.* 56:31–47.

Khan, R., and W. Rubin. 1975. Quantitation of submicrogram amounts of protein using Coomassie Brilliant Blue R on sodium dodecyl sulfate-polyacrylamide slab gels. *Anal. Biochem.* 67:347–352.

Kimura, M. 1968. Evolutionary rate at the molecular level. *Nature.* 217:624–626.

______. 1979. The neutral theory of molecular evolution. *Sci. Amer.* 241:98–126.

______. 1980. A simple method for estimating evolutionary rates of base substitutions through comparative studies of nucleotide sequences. *J. Molec. Evol.* 16:111–120.

King, E. E. 1970. Disc electrophoresis: Avoiding artifacts caused by persulfate. *J. Chromatog.* 53:559–563.

King, J. C., and T. Ohta. 1975. Polyallelic mutational equilibria. *Genetics* 79:681–691.

Kingsbury, N., and C. J. Masters. 1970. On the determination of component molecular weights in complex protein mixtures by means of disc electrophoresis. *Anal. Biochem.* 36:144–158.

Koen, A. L., and M. Goodman. 1969. Aconitase hydratase isozymes: Subcellular location, tissue distribution and possible subunit structure. *Biochim. Biophys. Acta* 191:698–701.

Koenig, R., and P. Gepts. 1989. Allozyme divesity in wild Phaseolus vulgaris: Further evidence for two major centers of genetic diversity. *Theor. Appl. Genet.* 78:809–817.

Kolin, A. 1953. An electromagnetokinetic phenomenon involving migration of neutral particles. *Science* 117:134–137.

Krebs, S. L., and J. F. Hancock. 1988. The consequences of inbreeding on fertility in *Vaccinium corymbosum. J. Amer. Soc. Hort. Sci.* 113:914–918.

______. 1989. Tetrasomic inheritance of isozyme markers in the highbush blueberry, *Vaccinium corymbosum* L. *Heredity* 63:11–18.

Kreitman, M. 1983. Nucleotide polymorphism at the alchohol dehydrogenase locus of *Drosophila melanogaster. Nature* 304:412–417.

Laemmli, V. K. 1970. Cleavage of structural proteins during the assembly of the head of bacteriophage T4. *Nature* 277:680–685.

Lai, L. Y. C. 1966. Variation of red cell acid phosphatase in two species of kangaroos. *Nature* 210:643.

Leaback, D. H., and A. C. Rutter. 1968. Polyacrylamide-isoelectric focusing. A new technique for the electrophoresis of proteins. *Biochem. Biophys. Res. Commun.* 32:447–453.

Lehninger, A. L. 1982. *Principles of Biochemistry.* New York: Worth Inc.

Lewis, W. H. P., and H. Harris. 1967. Human red cell peptidases. *Nature* 315–355.

Lewontin, R. C. 1974. *The Genetic Basis of Evolutionary Change.* New York: Columbia University Press.

Lin, C., G. Schipmann, W. A. Kittrel, and S. Ohno. 1969. The predominance of heterozygotes found in wild goldfish of Lake Erie at the gene locus for sorbital dehydrogenase. *Biochem. Genet.* 3:603.

L. K. B. Instruments (n.d.) Multiphor. *The Complete System for Electrophocusing, Electrophoresis and Isotachophoresis.* Washington: LKB.

Loening, V. E. 1967. The fractionation of high-molecular weight ribonucleic acid by polyacrylamide gel electrophoresis. *Biochem. J.* 102:251–257.

Longo, G. P., and J. G. Scandalios. 1969. Nuclear gene control of the mitochondrial malic dehydrogenase in maize. *Proc. Natl. Acad. Sci. USA* 62:104–111.

Loomis, W. D., and J. Battaile. 1965. Plant phenolic compounds and the isolation of plant enzymes. *Phytochemistry* 5:423–438.

MacDonald, T., and J. L. Brewbaker. 1974. Isozyme polmorphism in flowering plant. IX. The E5—E10 esterase loci of maize. *J. Hered.* 65:37–42.

______. 1975. Isozyme polymorphism in flowering plants. V. The isoesterases of maize: Tissue and substrate specificities and responses to chemical inhibitors. *Hawaii Agric. Exp. Sta. Tech. Bull.* 89:24.

Mahalanobis, P. C. 1936. On the generalized distance statistics. *Proc. Natl. Inst. Sci. India* 2:49–55.

Mahler, H. R., and E. H. Cordes. 1968. *Basic Biological Chemistry.* New York: Harper and Row.

Manicol, P. K. 1966. Peroxidase of the Alaska pea. *Arch. Biochem. Biophys.* 117:347–356.

Markert, C. L. 1957. Cited by Vallejos (1983). In *Isozymes in Plant Genetics and Breeding,* Part A. Eds. S. D. Tanksley and T. J. Orton. Amsterdam: Elsevier. 469–516.

______. 1968. Molecular basis for isozymes. *Am. NY Acad. Sci.* 151:753–763.

______. 1975. Biology of isozymes. *Bioscience* 25:365–368.

______. 1977. Isozymes: The development of a concept and its application. In *Isozymes: Current Topics in Biological and Medical Research.* New York: Allan R. Liss Inc. 1:1–17.

Markert, C. L., and F. Moller. 1959. Multiple forms of enzymes: Tissues, cytogenic and species specific patterns. *Proc. Nat. Acad. Sci. USA* 45:753–763.

Markert, C. L., and G. S. Whitt. 1968. Molecular varieties of isozymes. *Experentia* 24:977–991.

Marshall, D. R., and R. W. Allard. 1970. Maintenance of isozyme polymorphisms in natural populations of *Avena barbata. Genetics* 66:393–399.

Marshall, D. R., P. Broue, and A. J. Pryor. 1973. Adaptive significance of alcohol dehydrogenase isozymes in maize. *Nature* 244:16–17.

Mather, K. 1951. *The Measurement of Linkage in Heredity.* 2nd ed. London: Methuen and Co.

May, B., J. E. Wright, and M. Stoneking. 1979. Joint segregation of biochemical loci in *Salmonidae*: Results from experiments with *Salvelinus* and review of the literature on other species. *J. Fish. Res. Board Can.* 35:1114–1128.

McCord, J. M. 1979. Superoxide dismutase: Occurrence, structure, function and evolution. In *Isozymes: Current Topics in Biological and Medical Research.* New York: Allan R. Liss Inc. 3:1–21.

McCord, J. M., B. B. Keele, Jr., and I. Fredovich. 1971. An enzyme-based theory of obligate anaerobiosis: The physiological function of superoxide dismutase. *Proc. Nat. Acad. Sci. USA* 68:1024–1027.

McLeod, J. F. S. I. Guttman, and W. H. Eshbaugh. 1983. Peppers. In *Isozymes in Plant Genetics and Breeding,* Part B. Eds. S. D. Tanksley and T. J. Orton, Amsterdam: Elsevier. 189–202.

Meyer, T. S., and B. L. Lambert. 1965. Use of Coomassie brilliant blue R 250 for the electrophoresis of microgram quantities of parotid saliva proteins on acrylamide-gel strips. *Biochim. Biophys. Acta* 107:144–145.

Micales, J. A., M. R. Bonde, and G. L. Peterson. 1986. The use of isozyme analysis in fungal taxonomy and genetics. *Mycotaxon* 27:405–449.

Mitton, J. B., Y. B. Linhart, K. B. Sturgeon, and J. L. Hamrick. 1979. Allozyme polymorphisms detected in mature needle tissue of Ponderosa pine. *J. Hered.* 70:86–89.

Moore, G. A., and G. B. Collins. 1982. Identification of aneuploids in *Nicotiana tabaccum* by isozyme banding patterns. *Biochem. Genet.* 20:555–568.

______. 1983. Challenges confronting plant breeders. In *Isozymes in Plant Genetics and Breeding,* Part A. Eds. S. D. Tanksley and T. J. Orton. Amsterdam: Elsevier. 25–58.

Moran, G. F., and J. C. Bell. 1983. Eucalyptus. In *Isozymes in Plant Genetics and Breeding,* Part B. Eds. S. D. Tanksley and T. J. Orton. Amsterdam: Elsevier. 423–442.

Moss, B. W. 1982. *Isozymes.* Great Britain: Chapman and Hall.

Motojima, K., and K. Sakaguchi. 1982. Part of the lycyl residues in wheat alpha amylase is methylated to a N-E-trimethyl lysine. *Plant Cell Physiol.* 23:709–712.

Munkres, K. D., and F. M. Richards. 1965. Genetic alteration of *Neurospora* malate dehydrogenase. *Arch. Biochem. Biophys.* 109:457–465.

Nei, M. 1972. Genetic distance between populations. *Amer. Nat.* 106:283–292.

______. 1973. Analysis of gene diversity in subdivided populations. *Proc. Natl. Acad. Sci. USA* 70:3321–3323.

______. 1975. *Molecular Population Genetics and Evolution.* Amsterdam: North Holland.

Ni, W., E. F. Robertosn, and H. C. Reeves. 1987. Purification and charaterization of cytosolic NADP specific isocitrate dehydrogenase from *Pisim sativum Pl. Physiol.* 83:785–788.

Nichols, E. A., M. Chapman, and F. H. Ruddle. 1973. Polymorphism and linkage for mannose phosphate isomerase in *Mus musculus. Biochem. Genet.* 8:47–53.

Nichols, E. A., and F. H. Ruddle. 1973. A review of enzyme polymorphism, linkage and electrophoretic conditions for mouse and somatic hybrids in starch gels. *J. Histochem. Cytochem.* 21:1066–1081.

Nielsen, G. 1980. Identification of all genotypes in tetraploid rye grass (*Lolium* spp.) segregating for four alleles in a PGI-enzyme locus. *Hereditas* 92:49–52.

______. 1985. The use of isozymes as probes to identify and label plant varieties and cultivars. In *Isozymes: Current Topics in Biological and Medical Research.* 12:1–32. New York: Allan R. Liss.

Nijenhuis, B. T. 1971. Estimation of inbred seed in brussel sprout hybrid seed by acid phosphatase isoenzyme analysis. *Euphytica* 20:498–507.

Nishimura, M., and H. Beevers. 1979. Subcellular distribution of gluconeogentic enzymes in germinating castor bean endosperm. *Plant Physiol.* 64:31–37.

O'Farrell, P. H. 1975. High resolution two dimensional electrophoresis of proteins. *J. Biol. Chem.* 250:4007–4021.

Okunishi, M., K. Yamada, and K. Komagata. 1979. Electrophoretic comparison of enzymes from basidomycetes in different stages of development. *J. Gen. Appl. Microbiol.* 25:329–334.

O'Malley, D., N. C. Wheeler, and R. P. Guries. 1980. A manual for starch gel electrophoresis. Staff Paper Series No. 11. College of Agriculture and Life Sciences, University of Wisconsin.

Omenn, G. S., and P. T. W. Cohen. 1971. Electrophoretic methods for differentiation of glycolytic enzymes of mouse and human origin *In vitro* 7:132–139.

Ornstein, L. 1964. Disc electrophoresis. I. Background and theory. *Ann. NY Acad. Sci.* 121:321–349.

Orton, T. J. 1983. Celery and celeriac. In *Isozymes in Plant Genetics and Breeding,* Part B. Eds. S. D. Tanksley and T. J. Orton. Amsterdam: Elsevier. 339–350.

Parker, W. C., and A. G. Bearn. 1963. Boric acid induced heterogeneity of conalbumin starch gel electrophoresis. *Nature* 199:1184–1186.

Pearse, A. G. E. 1972. *Histochemistry. Theoretical and Applied.* Vol 2. 3rd ed. Baltimore: Williams and Wilkins.

Pierce, L. C., and J. L. Brewbaker. 1973. Applications of isozyme analysis in horticultural science. *HortScience* 8:17–22.

Prakash, S., R. C. Lewontin, and J. T. Hubby. 1969. A molecular approach to the study of genetic heterozygosity in natural populations. IV. Patterns of genetic variation in central marginal and isolated populations of Drosophila psuedoobscura. *Genetics* 61:841–858.

Quiros, C. F. 1981. Starch gel electrophoresis technique used with alfalfa and other *Medicago* species. *Can. J. Plant Sci.* 61:745–749.

______. 1982. Tetrasomic inheritance for multiple alleles in alfalfa. *Genetics* 101:117–127.

______. 1983. Alfalfa, luzerne. In *Isozymes in Plant Genetics and Breeding,* Part B. Eds. S. D. Tanksley and T. J. Orton. Amsterdam: Elsevier. 253–294.

Quiros, C. F., and K. Morgan. 1981. Peroxidase and Leucine-aminopeptidase in diploid *Medicago* species closely related to alfalfa: Multiple gene loci, multiple allelism and linkage. *Theor. Appl. Genet.* 60:221–228.

Racker, E. 1955. *Physiol. Rev.* 35:1–56.

Ramshaw, J. A. M., J. A. Coyne, and R. C. Lewontin. 1980. The sensitivity of gel electrophoresis as a detector of genetic variation. *Genetics* 93:1019–1037.

Raymond, S. 1964. Acrylamide gel electrophoresis. *Ann. NY Acad. Sci.* 121:350–365.

Raymond, S., and L. Weintraub. 1959. Acrylamide gel as a supporting medium for zone electrophoresis. *Science* 130:711.

Rennie, B. D., W. D. Beversdorf, and R. I. Buzzell. 1987. Genetic and linkage analysis of an aconitase hydratase variant in soybean (*Glycine max*). *J. Hered.* 78:323–326.

Ressler, N., E. Olivero, G. R. Thompson, and R. R. Joseph. 1966. Investigations of ribonuclease isozymes by an electrophoretic ultra-violet method. *Nature* 210:695–698.

Richmond, R. C. 1972. Enzyme variability in the *Drosophila willistoni* group. II. Amounts of variability in the superspecies *D. paulistorium. Genetics* 70:87–112.

Rick, C. M. 1976. Natural variability in wild species of *Lycospersicon* and its bearing on tomato breeding. *Genet. Agr.* 30:249–259.

______. 1983. Tomato. In *Isozymes in Plant Genetics and Breeding.* Eds. S. D. Tanksley and T. J. Orton. Amsterdam: Elsevier. 147–166.

Rick, C. M., J. F. Fobes, and S. D. Tanksley. 1979. Evolution of mating systems in *Lycopersicon hirsutum* as deduced from genetic variation in electrophoresis and morphological characters. *Plant Systematics and Evolution* 132:278–298.

Rick, C. M., and S. D. Tanksley. 1983. Isozyme monitoring of genetic variation in *Lycopersicon.* In *Isozymes: Current Topics in Biological and Medical Research.* 11:269–284. New York: Allan R. Liss.

Ridgway, G. T., S. W. Sherburne, and R. D. Lewis. 1970. Polymorphisms in the esterase of Atlantic herring. *Trans. Am. Fisheries. Soc.* 99:147–151.

Righetti, P. G., and J. W. Drysdale. 1974. Isoelectric focusing in gels. *J. Chromatog.* 98:271–321.

Robinson, K. 1966. An estimation of *Corynebacterium* spp. by gel electrophoresis. *J. Appl. Bacteriol.* 29:179–185.

Rodbard, D., and A. Chrambach. 1970. Unified theory for gel electrophoresis and gel filtration. *Proc. Nat. Acad. Sci. USA* 65:970–977.

Rogers, J. S. 1972. Measures of genetic similarity and genetic distance. *Studies in Genetics VII.* Univ. Texas Publ. 7213:283–292.

Salinas, J., and C. Benito. 1984. *Z. Pflanzenzuchtg.* 93:115–136.

Samaniengo, F. J., and P. Arus. 1983. On estimating the sib proportion in seed-purity determinations. *Biometrics* 39:563–572.

Scandalios, J. G. 1964. Tissue specific isozyme variation in maize. *J. Hered.* 55:282–285.

———. 1969. Genetic control of multiple molecular forms of enzymes in plants: A review. *Biochem. Genet.* 3:37–79.

———. 1967. Genetic control of alcohol dehydrogenase isozymes in maize. *Biochem. Genet.* 1:1–9.

———. 1974. Isozymes in development and differentiation. *Ann. Rev. Plant Physiol.* 25:225–258.

———. 1979. Controls of gene expression and enzyme differentiation. In *Physiological Genetics.* Ed. S. G. Scandalios. New York: Academic Press. 63–107.

Scandalios, J. G., J. C. Sorenson, and L. A. Ott. 1975. Genetic control and intercellular localization of glutamate oxaloacetate transaminase in maize. *Biochem. Genet.* 13:759–769.

Schnarrenberger, C., and A. Oeser. 1974. Two isoenzymes of glucosephosphate isomerase from spinach leaves and their intracellular compartmentalization. *Eur. J. Biochem.* 45:77–82.

Schwartz, D. 1960. Genetic studies on mutant enzymes in maize: Synthesis of hybrid enzymes by heterozygotes. *Proc. Nat. Acad. Sci. USA* 46:1210–1215.

———. 1969. Alcohol dehydrogenase in maize: Genetic basis for multiple isozymes. *Science* 164:585–586.

Schwartz, D., and W. J. Laughner. 1969. A molecular basis for heterosis. *Science* 166:626–627.

Schwartz, H. M., S. I. Biedron, M. M. von Holdtand, and S. Rehm. 1964. A study of some plant esterases. *Phytochemistry* 3:189–200.

Scopes, R. K. 1968. Methods for starch gel electrophoresis of sarcoplasmia. *Biochem. J.* 107:139–150.

Selander, R. K., M. H. Smith, S. Y. Yana, W. E. Johnson, and J. B. Gentry. 1971. Biochemical polymorphism and systematics in the genus *Peromyscus.* I: Variation in the old-field mouse (*Peromyscus polionotus*). *Studies in Genetics VI. Univ. Texas Publ.* 7103:49–90.

Shapiro, A. L., E. Vinuela, and J. V. Maizel. 1967. Molecular weight estimation of polypeptide chains by electrophoresis in SDS-polyacrylamide gels. *Biochem. Biophys. Res. Commun.* 28:815–820.

Shaw, C. R. 1964. *The Use of Genetic Variation in the Analysis of Isozyme Structure of Proteins: Biochemical and Genetic Aspects.* Upton, NY: Brookhaven National Laboratory.

———. 1965. Electrophoretic variation in enzymes. *Science* 149:936–943.

Shaw, C. R., and A. L. Koen. 1967. Aspartate dehydrogenase activity of malate dehydrogenase. *Biochim. Biophys. Acta.* 92:397–403.

———. 1968. Glucose-6-phosphate dehydrogenase and hexose 6-phosphate dehydrogenase of mammalian tissues. *Ann. NY Acad. Sci.* 151:149–154.

Shaw, C. R., and R. Prasad. 1970. Starch gel electrophoresis of enzymes: A compilation of recipes. *Biochem. Genet.* 4:297–320.

Shaw, D. V., and R. W. Allard. 1982. Estimation of outcrossing rates in Douglas-fir using isozyme markers. *Theor. Appl. Genet.* 62:11–120.

Sheen, S. J. 1972. Isozymic evidence bearing on the origin of *Nicotiana tabaccum* L. *Evolution* 26:143–154.

Shepherd, J. A., and L. Kalnitsky. 1954. Intercellular distribution of fumarase, aconitase and isocitric dehydrogenase in rabbit cerebral cortex. *J. Biol. Chem.* 207:605–611.

Shumaker, K. M., R. W. Allard, and A. L. Kahler. 1982. Cryptic variability at enzyme loci in three plant species: *Avena barbata, Hordeum vulgare,* and *Zea mays. J. Hered.* 73:86–90.

Siepman, R., and H. Stageman. 1967. Enzymeleck-trophorese in einschlu β-polymerisaten, phosphorylasen 2. A. Amylasen, phosphorylasen. *Z. Naturforsch.* 22b:949–955.

Simmonds, N. W. 1979. *Principles of Crop Improvement.* New York: Longmans. 170–173.

Sing, C. F., and G. J. Brewer. 1969. Isozymes of a polyploid series. *Genetics* 61:391–398.

Smith, E. E., and A. M. Rutenburg. 1966. Starch gel electrophoresis of human enzymes which hydrolyze L-leucine-β-naphthylamide. *Science* 152:1256–1257.

Smithies, O. 1955. Zone electrophoresis in starch gels. *Biochem. J.* 61:629–641.

Sokal, R. R., and F. J. Rohlf. 1969. *Biometry.* San Francisco, CA: W. H. Freeman.

Soltis, D. E., C. H. Hanfler, D. C. Darrow, and G. J. Gastony. 1983. Starch gel electrophoresis of ferns: A compilation of grinding buffers, gel and electrode buffers and staining schedules. *Amer. Fern J.* 73:9–27.

Soltis, D. E., and P. S. Soltis. 1988. Electrophoretic evidence for tetrasomic segregation in *Tolmiea menzieeii* (Saxifragaceae). *Heredity* 60:375–382.

———. Eds. 1989. *Isozymes in Plant Biology.* Portland, OR: Dioscorides Press.

Spencer, N., D. A. Hopkinson, and H. Harris. 1964. Phosphoglucomutase polymorphism in man. *Nature* 204:742–745.

———. 1968. Adenosine deaminase polymorphism in man. *Ann. Hum. Genet.* 32:9–14.

Sprecher, S. L., and E. C. Vallejos. 1989. System/tissue/enzyme combinations for starch gel electrophoresis of bean. *Annu. Rep. Bean Improv. Coop.* 32:32–33.

Staples, R. C., and M. A. Stahmann. 1963. Malate dehydrogenases in the rusted bean leaf. *Science* 140:1320–1321.

———. 1964. Changes in proteins and several enzymes in susceptible bean leaves after infection by the bean rust fungus. *Phytopathology* 54:760–746.

Stuber, C. W. 1989. Isozymes as markers for studying and manipulating quantitative traits. In *Isozymes in Plant Biology.* Eds. D. E. Soltis and P. S. Soltis. Portland, OR: Dioscorides Press. 206–220.

Stuber, C. W., R. H. Moll, M. M. Goodman, H. E. Schaffer, and B. S. Weir. 1980. Allozyme frequency changes associated with selection for increased grain yield in maize (Zea mays). *Genetics* 95:225–236.

Stuber, C. W., J. F. Wendel, M. M. Goodman, and J. S. C. Smith. 1988. Techniques and scoring procedures for starch gel electrophoresis of enzymes for maize (*Zea mays* L.). Tech. Bull. 286. NC Agric. Res. Station, NC State University.

Studier, F. W. 1973. Analysis of bacteriophage T7 early RNAs and proteins on slab gels. *J. Mol. Biol.* 79:237–248.

Suzuki, D. T., A. J. F. Griffiths, and R. C. Lewontin. 1981. *An Introduction to Genetic Analysis*. 2nd ed. San Fransisco, CA: W. H. Freeman. 21.

Svensson, H. 1961. *Acta Chem. Scand.* 15:325.

Takacs, O., and I. Kerese. 1984. Electrophoretic methods. In *Methods of Protein Analysis*. 1. Ed. I. Kerese. Chichester, UK: Ellis Horwood. 97–158.

Tanksley, S. D. 1979. Linkage, chromosomal association and expression and *Pgm-2* in tomato. *Biochem. Genet.* 17:1159–1167.

Tanksley S. D., and R. A. Jones. 1982. Application of alcohol dehydrogenase allozymes in testing the genetic purity of F_1 hybrids of tomato. *HortScience* 16:179–181.

Tanksley S. D., H. Medina-Filho, and C. M. Rick. 1982. Use of naturally occuring enzyme variation to detect and map genes controlling quantitative traits in an interspecific backcross of tomato. *Heredity* 49:11–25.

Tanksley, S. D., and C. M. Rick. 1980. Isozyme linkage map of the tomato: Applications in genetics and breeding. *Theor. Appl. Genet.* 57:161–170.

Thorpe, M. L., L. H. Duke, and W. D. Beversdorf. 1989. Procedures for the detection of isozymes of rapeseed (*Brassica napus* and *B. campestris*) by starch gel electrophoresis. University of Guelph Tech. Bull. TE OAC 887.

Thorup, O. A., W. B. Strole, and B. S. Leavell. 1961. A method for the location of catalase on starch gels. *J. Lab. Clin. Med.* 58:122–128.

Thurman, D. A., C. Palin, and M. V. Laycock. 1965. Isozymic nature of L-glutamic dehydrogenase of higher plants. *Nature* 207:193–194.

Ting, I. P., I. Fuhr, R. Curry, and W. C. Zschoche. 1975. Malate dehydrogenase isozymes in plants: Preparation, properties, and biological significance. *Isozymes* 2:369–384.

Trippa, G., A. Catamo, A. Lombardozzi, and R. Cicchetti. 1978. A simple approach for discovering common nonelectrophoretic enzyme variability: A heat denaturation study in *Drosophila melanogaster*. *Biochem. Genet.* 16:299–305.

Tyson, H., M. A. Fildes, and J. Starrobin. 1986. Genetic control of acid phosphatase R_m in flax (*Linum*) genotrophs. *Biochem. Genet.* 24:369–383.

Upcroft, J. A., and J. Done. 1974. Starch gel electrophoresis of plant NADH-nitrate reductase and nitrite reductase. *J. Exp. Bot.* 25:503–508.

Vallejos, E. C. 1983. Enzyme activity staining. In *Isozymes in Plant Genetics and Breeding*, Part A. Eds. S. D. Tanksley and T. J. Orton. Amsterdam: Elsevier. 469–513.

Vesterberg, O., and H. Svensson. 1966. *Acta Chem. Scand.* 20:820.

Vodkin, L. U., and J. G. Scandalios. 1981. Genetic control, developmental expression and biochemical properties of plant peptidase. In *Isozymes: Current Topics in Biology and Medical Research*. 5:1–25. New York: Allan R. Liss.

Wadstrom, T. 1974. Separation of Australia antigen and some bacterial proteins by isoelectric focusing in polyacrylamide gels. *Ann. NY Acad. Sci.* 209:405–413.

Wall, J. R. 1968. Leucine aminopeptidase polymorphism in *Phaseolus* and differential elimination of the donor parent genome in interspecific backcrosses. *Biochem. Genet.* 2:109–118.

Walter, H., F. W. Selby, and J. R. Francisco. 1965. Altered electrophoretic mobilities of some erythrocytic enzymes as a function of their age. *Nature* 208:76–77.

Ward, R. D., and D. O. F. Skibinski. 1988. Evidence that mitochondrial isozymes are genetically less variable than cytoplasmic isozymes. *Genet. Res.* 51:121–127.

Watts, R. L., and D. C. Watts. 1968. Gene duplication and the evolution of enzymes. *Nature* 217:1125–1130.

Weber, K., and M. Osborne. 1969. The reliability of molecular weight determination by dodecyl sulfate polyacrylamide gel electrophoresis. *J. Biol. Chem.* 244:4406–4412.

Weeden, N. F. 1983a. Evolution of plant isozymes. In *Isozymes in Plant Genetics and Breeding*, Part A. Eds. S. D. Tanksley and T. J. Orton. Amsterdam: Elsevier. 177–208.

______. 1983b. Isozyme variation at selected loci in *Pisum*. *Pisum Newslett.* 15:58–59.

______. 1989. Applications of isozymes in plant breeding. In *Plant Breeding Reviews*. Vol 6. Portland, OR: Timber Press. 11–53.

Weeden, N. F., and A. C. Emmo. n.d. *Horizontal Starch Gel Electrophoresis Laboratory Procedures*. New York Agricultural Experimental Station, Geneva, NY.

Weeden, N. F., M. M. Goodman, C. W. Stuber, and J. B. Beckett. 1988. New isozyme systems for maize (*Zea mays* L.): aconitate hydratase, adenylate kinase, NADH dehydrogenase, and shikimate dehydrogenase. *Biochem. Genet.* 26:739–748.

Weeden, N. F., and L. D. Gottlieb. 1979. Distinguishing allozymes and isozymes of phosphoglucoisomerases by electrophoretic comparisons of pollen and somatic tissues. *Biochemistry* 17:287–296.

______. 1980. The identification of cytoplasmic enzymes from pollen. *Plant Physiol.* 66:400–403.

Weeden, N. F., and J. F. Wendel. 1989. Genetics of plant isozymes. In *Isozymes in Plant Biology*. Eds. D. E. Soltis and P. S. Soltis. Portland, OR: Dioscorides Press. 46–72.

Wetter, L. R., and K. N. Kao. 1976. The use of isozymes in distinguishing the sexual and somatic hybridity in callus cultures derived from *Nicotiana*. *Z. Plazanphysiol.* 80:455–462.

Whightman, F., and J. C. Forrest. 1978. Properties of plant amino transferases. *Phytochemistry* 17:1455–1471.

Wijsma, H. F. W. 1983. Petunia. In *Isozymes in Plant Genetics and Breeding*, Part B. Eds. S. D. Tanksley and T. J. Orton. Amsterdam: Elsivier. 229–252.

Wilkinson, J. H. 1970. Techniques for the separation of isozymes. In *Isoenzymes*. Philadelphia: Lippincott.

Wilson, R. E., and J. F. Hancock, Jr. 1978. Comparison of four techniques used in the extraction of plant enzyme for electrophoresis. *Bull. Torrey Bot. Club* 105:318–320.

Worthington Manual. 1968. Enzymes. Worthington Biochem. Corp.

Wright, D. A., M. J. Siciliano, and J. N. Baptist. 1972. Genetic evidence for the tetramer structure of glyceraldehyde-3-phosphate dehydrogenase. *Experientia* 28:889.

Wright, S. 1951. The genetical structure of populations. *Ann. Eugen.* 15:323–354.

Zubay, J. 1983. *Biochemistry*. 2nd ed. New York: MacMillan.

SUBJECT INDEX